Around the World

VOLUME TWO

CONTENTS

Im Auftrag des Verlages erarbeitet von:
Wolfgang Biederstädt, Köln
Karen Carlson-Kreibohm, Berlin
Dieter Haupt, Wolfenbüttel

Redaktion: Marc Proulx

Redaktionsassistenz: Silke Berger, Bettina Schaschke

Beratende Mitwirkung: Susan Richards, Cambridge

Technische Umsetzung: Kai Reddig

Umschlagfotos: Verlagsarchiv

Bildquellen: siehe Verzeichnis auf Seite 80

Teile dieses Materials basieren auf der Lehrwerksreihe *Mensch und Raum · Geographie*. Wir danken den Autoren der Cornelsen Redaktionsgesellschaft, Laatzen, für die freundliche Erlaubnis zur Verwendung.

Cornelsen online http://www.cornelsen.de

1. Auflage € Druck 6 5 4 3 Jahr 05 04 03 02

Alle Drucke dieser Auflage können im Unterricht nebeneinander verwendet werden.

Druck: CS-Druck Cornelsen Stürtz, Berlin

ISBN 3-464-07185-5

Bestellnummer 71855

Gedruckt auf säurefreiem Papier, umweltschonend hergestellt
aus chlorfrei gebleichten Faserstoffen.

......The United States of America

DID YOU KNOW ...?

Washington, DC, the capital, is not in any state. It is a separate "District of Columbia" located in the middle of the first 13 states.

The flag of the United States has 50 stars, one for each of the 50 states. The 13 stripes stand for the first 13 states.

The area of the United States is 27 times larger than Germany, but the population is only about three times greater.

Alaska and Hawaii, the two newest states, are separated from the continental United States.

The five Great Lakes between the USA and Canada form the greatest connected area of fresh water on Earth.

At Mount Waialeale in Hawaii it rains 350 days a year. The sunniest spot in the USA is Yuma, Arizona, with 330 sunny days a year.

The world's three busiest airports are all in American cities: Atlanta, Chicago, and Los Angeles.

More people live in California than in any other state: over 30 million. More than 10 per cent of all Americans live there.

Over 75 per cent of the people in the United States live in cities: the largest ones are New York, Los Angeles and Chicago.

Figure 1 Climates and topography

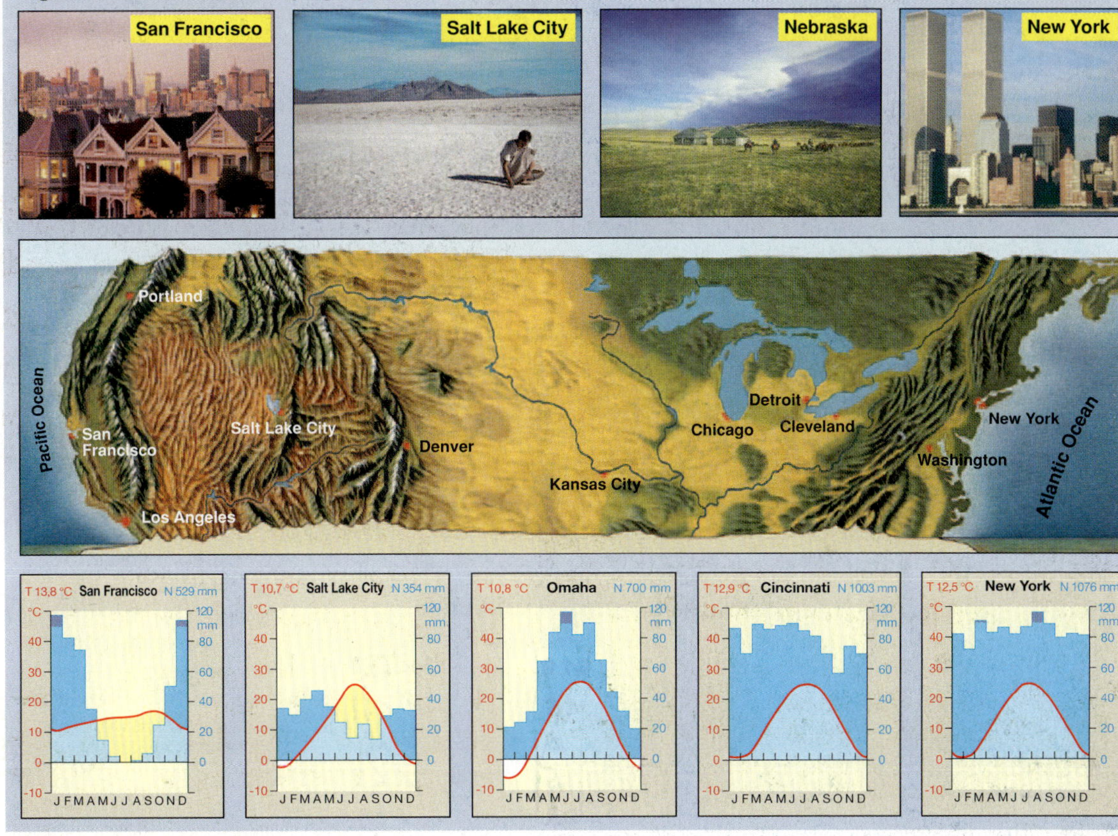

GEOGRAPHY TASKS

1 On a map of the United States locate the cities and states mentioned on page 2. What other places in the USA do you know from television, travelling, penfriends, etc.?

2 Using the cross-section map on this page and your atlas, draw an outline sketch (east-west) showing the main relief features of the USA.

3 Work out the approximate distances from the Canadian to the Mexican border and from the East Coast to the West Coast. About how long do you think it would take to drive from New York to Los Angeles?

4 Find out the latitude of Frankfurt/Main and find this line of latitude on the North American continent. What do you notice about its location?

5 Compare the climates of the West Coast and the East Coast, as shown in the climate graphs for San Francisco and New York.

6 Rome in Italy is on the same latitude as New York City. In what ways is the climate of Rome different from that of New York?

7 Draw a climate graph of location (A) with the following statistics.

	Jan	Feb	Mar	Apr	May	Jun	Jul	Aug	Sep	Oct	Nov	Dec	Year
°C	4.7	5.9	7.6	10.3	13.3	15.8	18.1	17.9	15.4	11.7	8.0	5.9	11.2
mm	126	98	81	56	44	35	16	18	43	76	125	139	825

Where in the USA do you think location (A) might be? Explain your answer.

PEOPLE IN THE USA

Everyone in the USA has ancestors who originally came from somewhere else. Even the Native Americans, often called Indians because Columbus thought he was near India, came from Asia many thousands of years ago. During the 1800s many immigrants came from Germany and other European countries. Today several thousand people still move to America every year from every place on Earth. About 10 per cent of all Americans were born in another country. One of the fastest growing groups is the Hispanics, Spanish-speaking people from Mexico and other Latin American countries.

Figure 2 **Population change in the USA by ethnic group**

	1980		2000		2040*	
	Million	%	Million	%	Million	%
Whites	180.9	79.9	211.5	75.1	287.8	~~76.3~~ 62.8
Blacks	26.1	11.5	34.7	13.0	54.5	~~14.4~~ 11.9
Hispanics	14.6	6.4	35.3	13.0	82.7	~~21.9~~ 18.0
Asians	3.6	1.6	10.2	3.6	29.5	~~7.8~~ 6.4
Native Americans	1.3	0.6	2.4	0.9	4.0	~~1.1~~ 0.9
Total	**226.5**	**100.0**	281,4	**100.0**	~~**377.3**~~ 458,5	**100.0**

*estimate

Figure 3 **Newcomers to the USA**

	Country	Immigrants each year
1	Mexico	173,900
2	China, People's Republic	45,700
3	Philippines	42,500
4	India	42,000
5	Vietnam	26,700
6	Nicaragua	24,000
7	El Salvador	22,600
8	Haiti	22,400
9	Cuba	20,800
10	Dominican Republic	17,500
11	Russia	17,100
12	Canada	16,200
13	Jamaica	16,000
14	Korea	15,830
15	Ukraine	15,810
16	Pakistan	14,530
17	Colombia	14,500
18	United Kingdom	13,400
19	Bosnia-Herzegovina	11,800
20	Poland	10,800
21	Guatemala	10,000
22	Peru	9600
23	Taiwan	9000
24	Nigeria	7900
25	Germany	7600
26	Bangladesh	7215

Source: US Immigration and Naturalization Service, 2001

PRACTISE YOUR ENGLISH

You can use these structures when discussing demographic data:

The number of white/Hispanic Americans will probably increase/decrease by ...% between the years 2000 and 2040.

It is predicted that ... / According to demographic predictions ... / If present trends continue, there will be ...

The fastest growing ethnic group is the ...

In 1980 there were 26.1 ("twenty-six point one") million / 26,100,000 ("twenty-six million, one hundred thousand") whites.

GEOGRAPHY TASKS

1 **Using the statistics from Figure 3, list the immigrants by region (Europe, Africa, Asia, Latin America / Caribbean, Canada) and find the total numbers that have come from each region. What reasons do you think these people might have for immigrating to America?**

2 **Draw line graphs to show the population changes in each of the five ethnic groups between 1980 and 2040. What main changes are expected?**

A LOOK AT THE USA AND ITS ECONOMY

Figure 4 **Comparative data (2001)**

	USA	Russia	European Union*	Germany
Area (km²)	9,809,155	17,075,400	3,234,000	357.022
Population (millions)	281,422	144,300	370.3	82,183
Population density (person per km²)	28,7	8.5	114.5	230.2
Life expectancy (in years)	76 (m 75/w 80)	66 (m 60/w 72)	77 (m 74/w 80)	78 (m 75/w 81)
Gross domestic product (mil US $)	9,872,9	246	7,471,620	3,976
GDP per person (US $)	36,000	6800	20,177	24,900
% of GDP from agriculture	1.3	7.0	–	1.0
% of GDP from manufacturing	24.0	38.0	–	28.0
% of GDP from services	74.7	55.0	–	71.0
Energy consumption (kg oil/units per person)	2180	608	–	331

*Statistics from 1993-1994

Source: Spiegel Almanach, 2002

PRACTISE YOUR ENGLISH

The following structures can be used when describing and comparing data:

The USA is about … times larger than … .
There are twice / three times more people in … than in … .
The population/area of … is x times greater than that of … .
The population density is about 27 persons / 27.5 ("twenty-seven point five") persons per square km.
The primary/secondary/tertiary sector accounts for … % of the gross domestic product.

Figure 5 **Contribution of different industries to the gross domestic product (GDP) of the USA (in %)**

	1960	1996	2000
Agriculture, forestry, farming	4.2	1.7	1.4
Mining	2.6	1.4	1.3
Construction	4.6	4.0	4.7
Manufacturing	28.6	16.8	15.9
Transportation, communication	9.1	8.5	8.4
Trade	16.7	15.6	15.9
Finance, banking, insurance	14.3	18.4	19.6
Service	9.7	20.0	21.9
Government	10.5	13.1	12.3

GEOGRAPHY TASKS

1 **Using the statistics in Figure 4, compare the USA, Russia, the European Union and Germany in terms of area, population and economic data.**

2 **Draw pie graphs for the USA, Russia, and Germany showing the relative amount of the GDP contributed by the primary (agriculture), secondary (manufacturing) and tertiary (service) sectors of the economy.**

3 **List the categories in Figure 5 according to the amount of change since 1960, beginning with the category with the greatest increase.**

4 **Give examples of jobs in each of the fields in Figure 5. Which categories of work are considered to be in the primary/secondary/tertiary sectors? In which sector/field of work do you think you will work someday?**

FARMING IN THE USA

Anyone who has driven in the central part of the USA and seen nothing but cornfields for hours and hours will not be surprised to hear that America produces more corn (as maize is called in North America) than any other country in the world: statistically about five pounds per day per person in the USA. Of course only a small amount is eaten directly as corn-on-the-cob or popcorn. About three-quarters of it goes to feed animals, mostly pigs, and is later eaten in the form of pork or bacon.

The area where most of the corn grows – in Iowa, Illinois and Indiana – is sometimes called the Corn Belt, although other crops and animals are also raised there. The Dairy Belt, where much of the nation's milk and cheese is produced, stretches across the northern part of the USA from Minnesota to Maine. West of the Dairy and Corn Belts is an area between Texas and North Dakota where most of the wheat is grown. In the days of plantations and slaves, "King Cotton" ruled the south-eastern states, but today, with modern irrigation methods, cotton grows better in California and Texas than in the old Cotton Belt of the South.

Except for coffee, tea, cocoa, bananas and a few other tropical products, America grows every crop it needs to feed its people as well as millions of others in the world. America is by far the world's largest exporter of agricultural products. In world production statistics, America is the leading producer of corn, soybeans, beef, milk, cheese and chicken, and the second highest producer of cotton.

Typical for much of the farmland in the USA is the checkerboard pattern of large square fields separated by long straight roads. This goes back to the days when the government systematically divided up the land that had been Indian territory and sold it to new immigrants and Americans from the East Coast. The standard-size farm was 160 acres, about 65 hectares, more than most immigrants could ever dream of owning in Europe.

But today, a 160-acre farm is seldom large enough to provide enough income for a farm family unless they specialize in something unusual or have second jobs outside of the farm. In order to survive, most farmers have to buy or rent much more land. This is especially difficult when farmers also need to buy expensive machines and when prices they get for their crops and animals are low. The 1980s was an especially hard time for agriculture in America and thousands of farmers had to give up and look for other work.

Figure 6 **Agricultural regions of the USA**

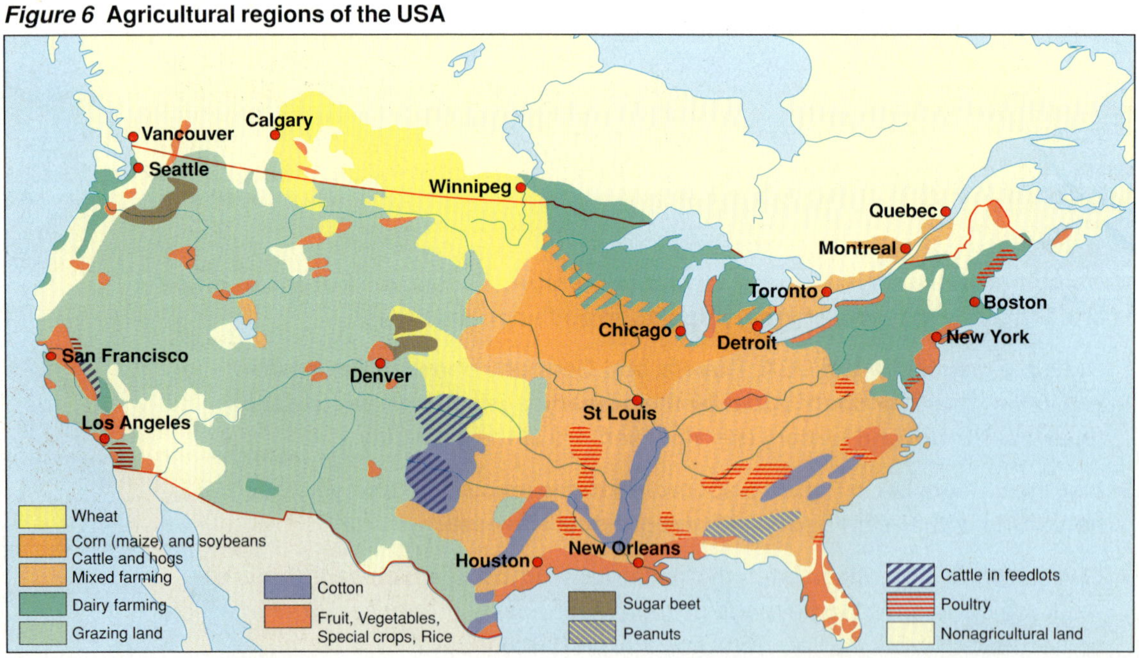

Many agricultural tools and machines were first made in America and farms there were mechanized earlier than those in other countries. One advantage of course is that fewer workers are needed and much of the work is easier. A disadvantage is that the new machines are very expensive. In order to make the investment worthwhile, farmers need to have bigger and bigger farms and specialize in large areas of the same crop. Most farmers who raise animals also believe that "more is better". It is not uncommon for thousands of pigs, cattle, chickens, or turkeys to live in a relatively small space so the energy in their food can go directly to growth and is not "wasted" on movement.

Just as some farmers are specializing in one or two crops or kinds of animals, others specialize in selling a service to other farmers, for example the owners of small airplanes that are used to plant seeds and spread fertilizer or pesticides from the air. About 10 per cent of the crops in America are grown with the help of these "crop dusters", as the little planes are called. Another new field is in high-tech agricultural equipment, including computers on fertilizer spreaders that can connect to satellites and tell the machines exactly how much fertilizer to spread in which place.

But even with high-tech equipment, there is still a lot of farm work that needs to be done by hand, especially in the fruit and vegetable industry. Thousands of migrant farmworkers are needed along the east and west coasts as well as in Texas and in central areas. Over 85 per cent of these people are from minority groups, most are Spanish-speaking people from Mexico and other Latin American countries. Even though they do not earn very much money, and the living and working conditions are sometimes not very good, they know they can earn much more in America than in their poorer home countries.

Figure 7 **Changes in agriculture in the USA**

	Number of farms (millions)	Number of people (millions)	Average farm size (ha)	Tractors (millions)	Use of chem. fertilizers (million tons)	Corn (maize) production (100 kg/ha)
1930	6.5	12.4	57	1.1	5.5	16
1960	3.9	7.0	119	4.6	37.1	42
1992	2.1	2.7	190	4.8	18.7	83

GEOGRAPHY TASKS

1. **Compare the map in Figure 6 with maps showing relief and climate features of the USA. Describe the geographical regions in which different agricultural products are grown.**
2. **Describe the main changes in agriculture in the USA shown in Figure 7.**
3. **Find out about changes in agriculture in your region during the last 50 years. Pick out from the text three ways in which farming in the USA has become more high-tech. How do these changes compare with those in your region?**

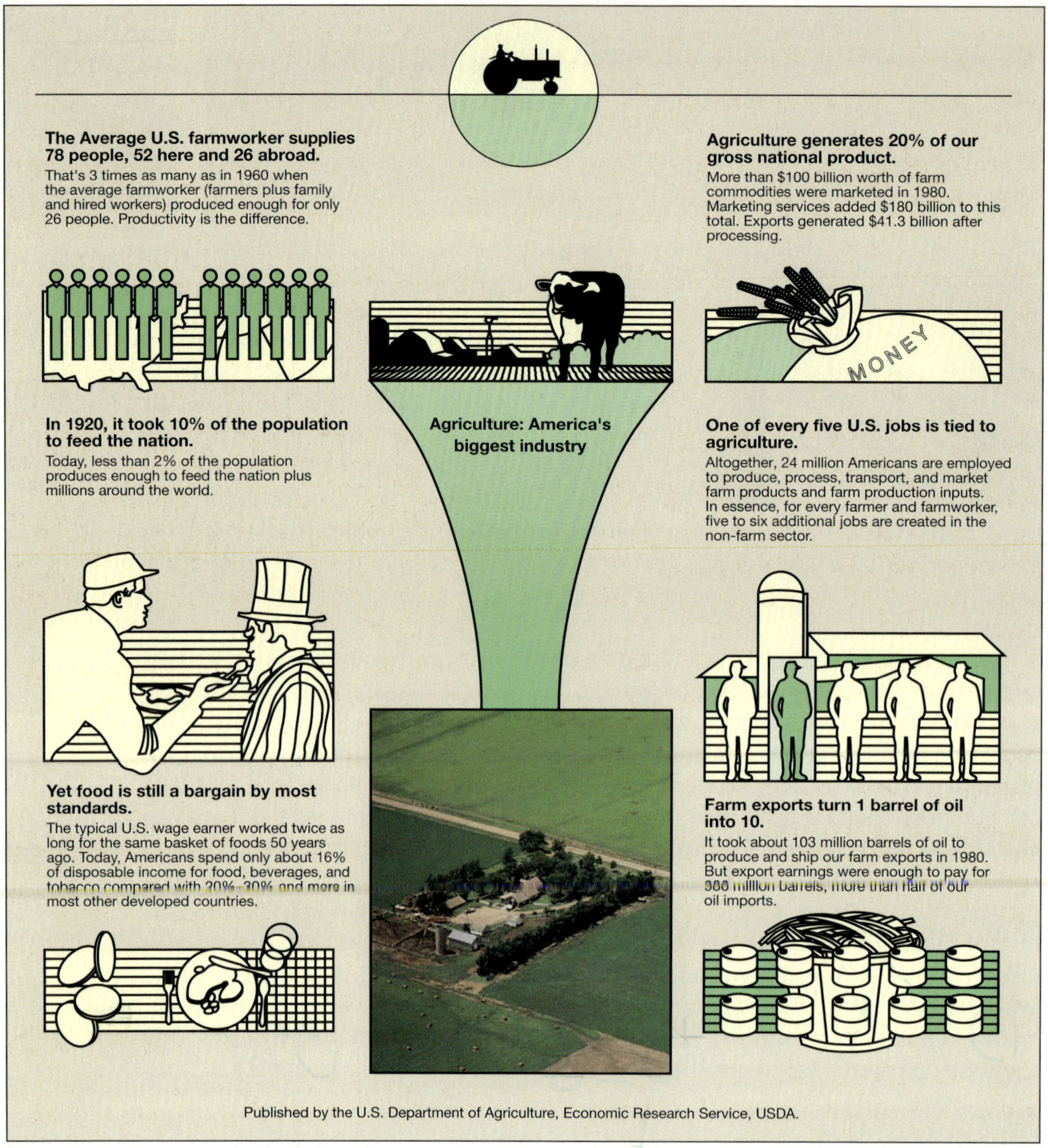

Figure 8 **Agriculture in the USA**

THE BEEF BUSINESS

Americans eat over five billion hamburgers every year. Producing beef for hamburgers and steaks is a big business. In fact, the cattle industry creates more income than any other part of American agriculture. There are over a million cattle farmers and ranchers in the USA and cattle are raised in all 50 states. The leading cattle states are Texas, Kansas, Nebraska, Oklahoma and California.

The USA produces 25 per cent of the world's beef, although only ten per cent of all the cattle in the world are in the USA. This means that beef production is very efficient. Cattle produce much more beef per animal than in the past. The beef business uses automation in feeding the animals and processing the meat as well as genetic technology to improve the health and productivity of the animals. The cattle business is becoming more concentrated with fewer farms and ranches and more animals per farm.

Feedlots are an important part of the beef industry. These are large fenced areas where young cattle are brought for a few months to be fattened or "finished", as it is called, before being slaughtered. Some feedlots have only a few hundred animals, others over 100,000. In cold climates they are covered, in the western states they are usually open. Automated feeding trucks go through the feedlots and give the animals the exact amount of special concentrated feed they need until they weigh from 900 to 1400 pounds each and are about two years old.

Most cattlemen specialize in only one part of the business. Some only raise mother cows and their young calves until the calves are old enough to leave their mother, at about seven months. Sometimes calves are then sold directly to a feedlot, others first spend a few months with another cattleman on grazing land. Cattle can graze on land that is not suitable for growing crops.

While most of the cattle in the USA are raised on small or medium-sized family farms or ranches, part of the industry is owned and managed by large corporations. In some cases, a large holding company manages several other businesses that include every part of the beef industry such as farms that grow feed grain, feedlots, companies that make automatic feeding machines, transportation firms, meat-packers, and even restaurants and supermarkets.

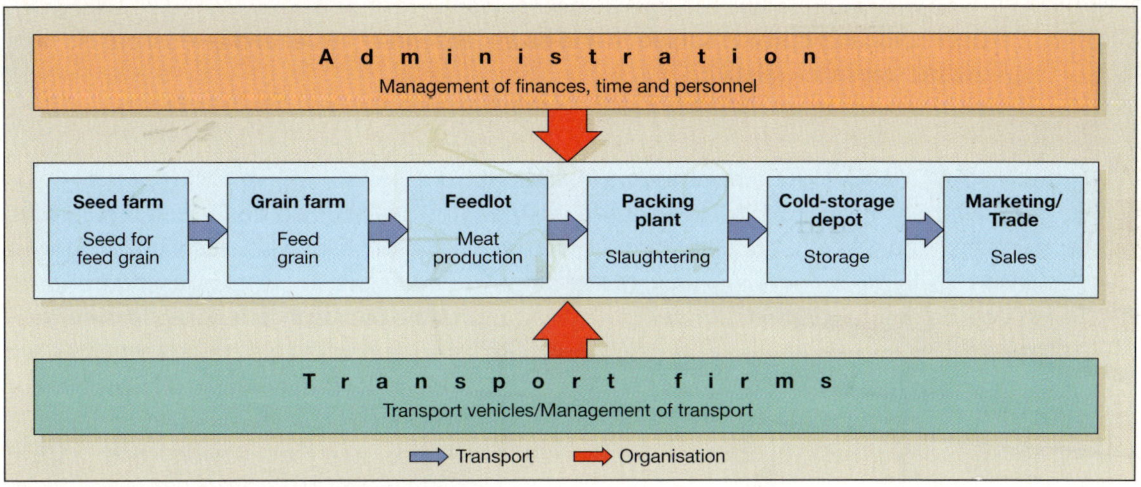

Figure 9 **Organisation of a holding company**

GEOGRAPHY TASKS

1. **Describe what a feedlot looks like.**
2. **The text describes a huge increase in the productivity of beef farming. List the developments that have made beef farming more productive.**
3. **The beef business is best described as an industry. In what ways is it similar to a manufacturing industry?**
4. **Why do some people object to the beef farming methods described in the text?**

VOICES FROM THE HEARTLAND

RICK TWEDT – Decorah, Iowa

My great-grandfather came from Norway and bought a 171 acre farm (about 70 ha) in central Iowa. In those days they raised all kinds of animals – horses, cattle, hogs, sheep, goats, chickens as well as several different crops – corn, soybeans, oats, alfalfa. We are proud that this farm is still in the family after over a hundred years, but all that is left of it is 40 acres. My mom and dad, who are 70 and 73, still live there and raise a few chickens, ducks and geese, but more as a hobby. My uncles had to sell their parts of the 171 acres in the 1980s and now that land is part of a huge farm where they grow only corn and soybeans.

The sentimental side of me longs for the old-fashioned small family farm with fewer machines and more people, but I know we can't really go back to those days. I teach agriculture classes at Decorah High School and while I know that my students will never be farmers like my parents or grandparents, there are so many new and exciting opportunities for them in new fields of high-tech equipment, genetics and farm management. In some ways, I think people today have more respect for the land and know that we have to take care of it in order to feed our hungry planet. There is a trend to using fewer chemicals, for example, and only where it is really necessary.

As the youngest son, I could take over our family farm someday. Will I do it? Not likely. But I'll probably rent it to someone and be an "absentee" owner and at least our old farm will still stay in the family.

NANCY HASTING – Mount Vernon, Indiana

We have quite a bit of land here in southern Indiana, but we don't farm it ourselves anymore. We rent it to other farmers who do the actual farmwork, but my husband Mike manages it. He also spends a lot of time putting in a new irrigation system, so that when we have a dry year, we don't lose everything. My part in our business is a greenhouse operation. I raise and sell flowers and garden plants. And of course I keep busy with our two children, Daniel, 10, and Diane, 5.

For the most part, people today need to farm larger and larger pieces of land in order to make a profit. The equipment has become so expensive; a new combine costs more than a really nice large house. A lot of farmers are also trying out new crops. Several around here are starting to raise ostrich for meat ("tastes like beef, as healthy as chicken"), feathers and leather (popular with fashion designers!).

WELCOME TO THE WHEATFIELDS

Bill and Marla Ashmore are the owners of "A-Team Custom Harvesting". Together with their sons Justin, 20, and Jim, 16, their crew of nephews and friends of the boys, and their four combines and five trucks, they harvest wheat, corn and sunflowers from Oklahoma to Nebraska.

Marla reports:

When my husband Bill and I were first married, we wanted to farm but we had no family financial backing. We decided to buy a combine for ourselves and make some money by using it to cut wheat for other farmers, too. Our first crew was my husband and myself and our two boys who were still babies. I drove the truck and cared for the kids, while my husband drove a very noisy old combine and did all the repairs. There was no air-conditioning in those days and the children were forever dirty, hot and tired. But it was our life. The boys started running combines themselves when they were ten years old. Today we still live on our farm in Kansas (with two dogs, some cats and one rooster) and we cut for about 25 other farmers who either don't have time to do it themselves or the money to buy a $150,000 combine.

During the winter and spring Bill is busy buying, selling and repairing machines and equipment (I get to wash all those greasy clothes!) and I work with handicapped children at school. We're all glad when it's the end of May and time to leave for our first harvesting job in Oklahoma because that's when the fun begins. Don't get me wrong, it's still hard work, we usually work 15–18 hours a day – sometimes 24 – but custom cutting gets in your blood. There's something special about pulling into a ripe field and harvesting it, especially since we're farmers ourselves and know how much work it takes to get a field that far. And even if the crew gets on each other's nerves sometimes, for the most part we like being around each other and develop a special closeness.

We sleep in cheap motels along our route. We have a big hot breakfast at a cafe in the morning, make sandwiches for lunch, and then I cook a hot meal for the crew late in the evening. When the guys cook, they often have diesel dogs – that's what they call hot dogs if they put too much diesel on the fire. We work our way north through Kansas, part of Colorado and finish in Sidney, Nebraska, around the first of August. Then it's back home to Kansas for a short vacation before the new school year begins for me and the boys.

GEOGRAPHY TASKS

1 **The A-Team travels from southern Oklahoma to western Nebraska. Follow their route on a map and measure about how far that is.**

2 **Some harvesting crews begin in Texas in May and finish in North Dakota in September. Explain why.**

3 **After reading the reports on pages 10 and 11, explain how Rick, Nancy and Marla have each adapted to the changing situation of the farm industry.**

AGRIBUSINESS IN CALIFORNIA

"It never rains in Southern California" is only the first part of the song. When it rains there, "it pours, man, it pours". There is plenty of water in California but it is in the "wrong places" and it comes at the "wrong time". Some of the world's best farmland is in California's long Central Valley, but it is very dry and needs water that comes through a huge system of reservoirs, dams, canals and aqueducts. Some of the water travels over 800 kilometres.

With water almost anything will grow in California, which is by far the leading agricultural state in the USA, producing twice as much as Texas, the second leading state. Over 250 different crops are grown in California, everything from artichokes and avocados to watermelons and zucchinis. Although less than a third of the land is used for farming, California produces half of all the fruit and vegetables grown in the USA. About half of California's agricultural products are exported, mostly to Japan and other Asian countries, with 10 per cent going to the European Union.

California is also a leader in the trend toward bigger and bigger farms and more and more technology. Biologists at California universities work on creating "improved" plants, such as tomatoes with thicker skins and a cube-like shape, which can be picked and packed easier by machine. Much of the land in California is owned by huge agribusiness corporations that also own and operate food-processing factories as well as packing, shipping and marketing companies.

Value in 1996 (in millions of $)

1	Milk and cream	3717	11	Strawberries	585
2	Grapes	2159	12	Oranges	532
3	Nursery products	1610	13	Chickens	458
4	Cattle and calves	1151	14	Broccoli	369
5	Cotton	1074	15	Eggs	367
6	Almonds	1009	16	Walnuts	322
7	Hay	813	17	Rice	285
8	Lettuce	735	18	Carrots	268
9	Flowers	703	19	Wheat	258
10	Tomatoes	664	20	Peaches	249

Figure 10
California's top 20 farm products

Value in 1996 (in millions of $)

1	Beef products	993	6	Oranges	335
2	Cotton	799	7	Dairy	236
3	Grapes & products	674	8	Lettuce	154
4	Almonds	488	9	Prunes	129
5	Fish/seafood	440	10	Walnuts	125

Figure 11
California's top 10 export products

GEOGRAPHY TASKS ■■■

1 **Find out which of California's top 20 agricultural products are also grown in Germany. Explain reasons for the differences.**

2 **Find out what California products are sold in your local supermarket. Check the labels and locate the places of origin on a map of California.**

3 **Work out the percentage of California's production of cotton, almonds, oranges, lettuce and walnuts that is exported. Do this by comparing the values for production and export in Figures 10 and 11. Which of these five crops is grown mainly for export?**

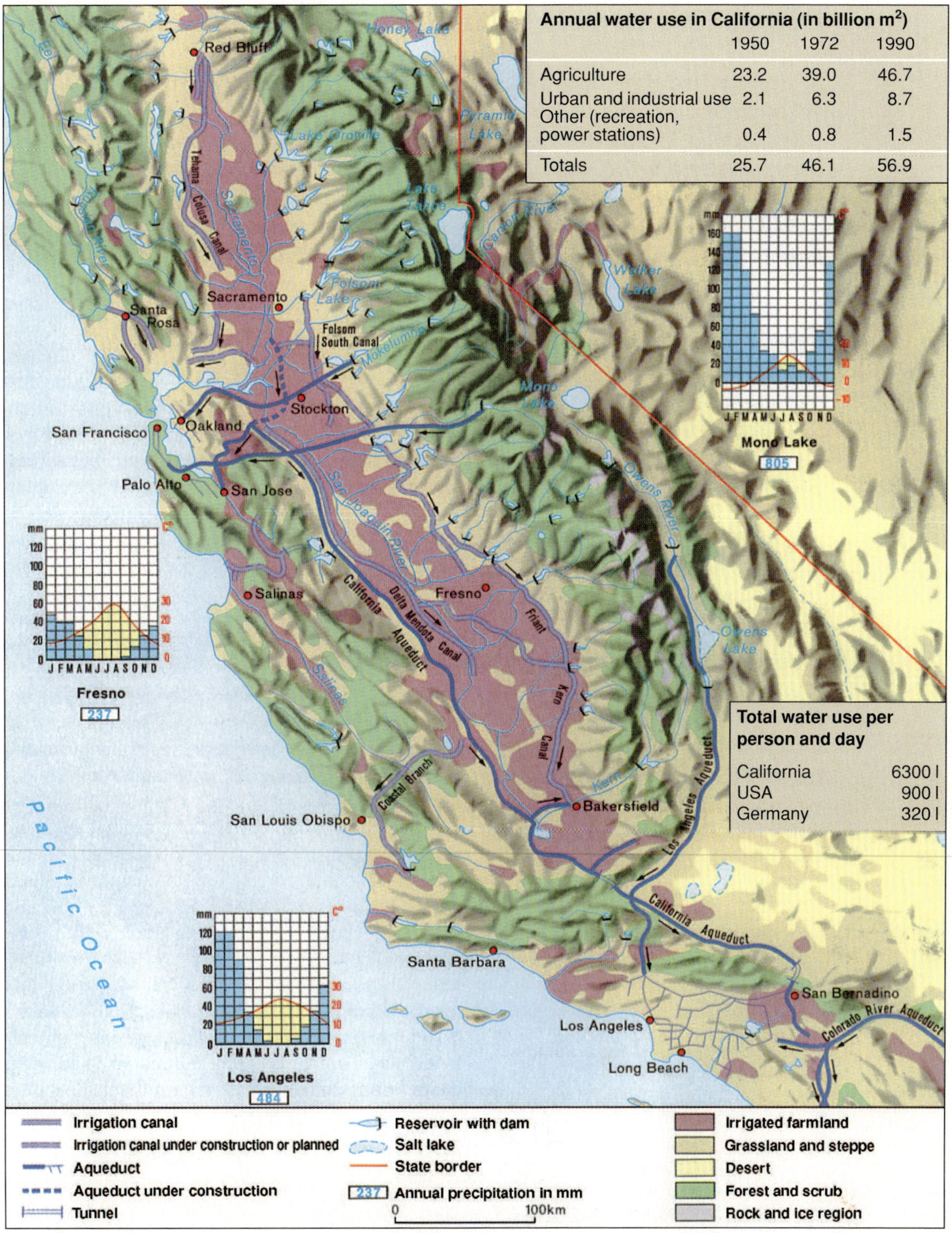

Annual water use in California (in billion m²)

	1950	1972	1990
Agriculture	23.2	39.0	46.7
Urban and industrial use	2.1	6.3	8.7
Other (recreation, power stations)	0.4	0.8	1.5
Totals	25.7	46.1	56.9

Total water use per person and day

California	6300 l
USA	900 l
Germany	320 l

4 **Draw a simplified sketch map of California showing the main land use features.**

5 **Use the climate graphs to compare the climate of Los Angeles and Mono Lake.**

6 **Explain the location of California's reservoirs and dams by referring to the information on relief and climate.**

7 **Explain the statement: "The main problem of California agriculture is not that there is not enough water, but that it is not in the right places."**

INDUSTRY IN THE SNOW BELT AND THE SUN BELT

At the beginning of the twentieth century, the United States had grown to be the greatest industrial nation in the world and its Manufacturing Belt had become the world's largest industrial area. This centre of industry stretches from the north Atlantic coast to the Great Lakes with Boston, Baltimore, Milwaukee and St Louis marking the corners of a region that also includes the cities of Chicago, Detroit, Pittsburgh and Cleveland.

Heavy industry depends on iron ore, coal and transportation. Iron ore was found in the Lake Superior area and could be transported across the Great Lakes to cities closer to the coal found in the Appalachian Mountains. Huge numbers of immigrants as well as American farm workers from the South who were no longer needed in the fields were a ready source of workers in the factories which were producing steel for the railroads, bridges and skyscrapers of America, as well as making cars, ships and machines for the growing nation.

Today, the top two manufacturing states are California and Texas, places that are far away from the old Manufacturing Belt, which is now often called the Snow Belt (or Frost Belt) because of its cold winters, or the Rust Belt because the traditional steel and coal industries are "rusty" compared to the newer industries of electronics, petrochemicals and information technology. These growth industries are especially strong in the area south of 37° latitude, a region commonly called the Sun Belt.

One of the first industries to move south and west was the film industry which could use the California sun for lighting and make films outside all year round. Clear skies and wide open spaces were also important to the aircraft and space industries. During World War II the Pacific coast and the Gulf region were important for strategic military reasons and the US government invested heavily in defence industries in those areas. Large amounts of oil and natural gas as well as nuclear power plants are sources of energy found in the South and the West. Many firms are attracted to the Sun Belt because in general, land is cheaper and taxes are lower. Labour costs are also often lower because unions are not so strong. There is no problem finding enough workers, because thousands of people are glad to get away from the cold winters of the Snow Belt.

Figure 13 **Natural resources and industries in the Manufacturing Belt**

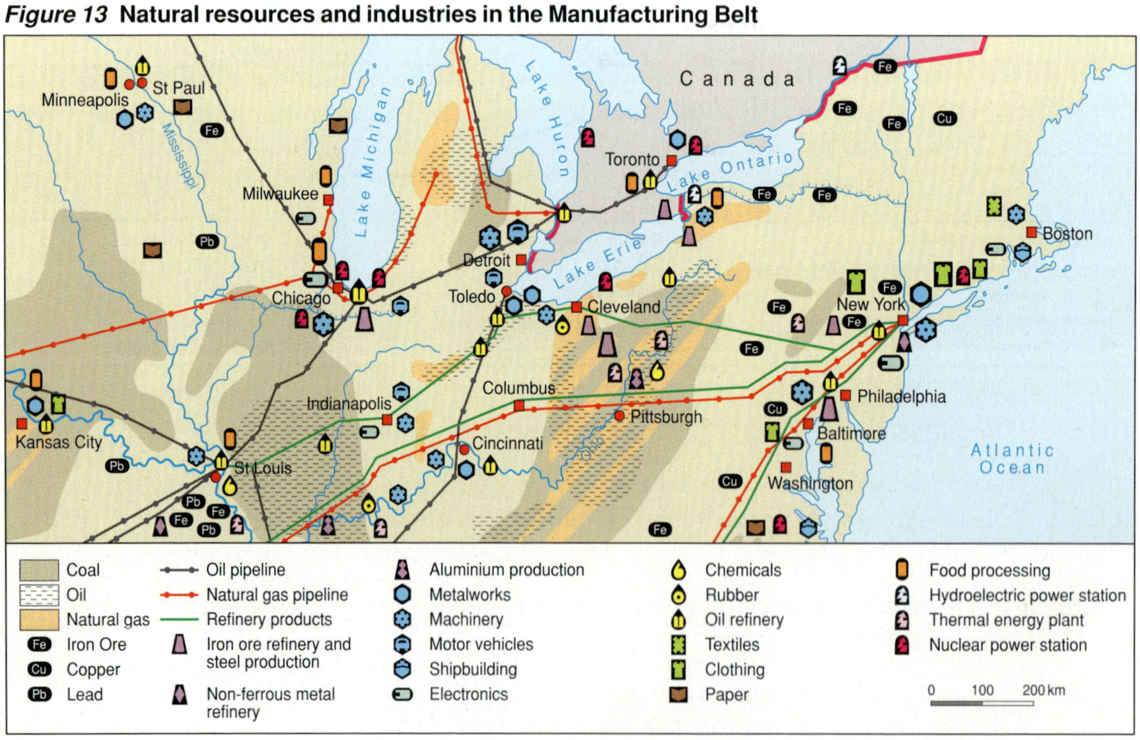

Figure 14 **Changing patterns in the industrial regions**

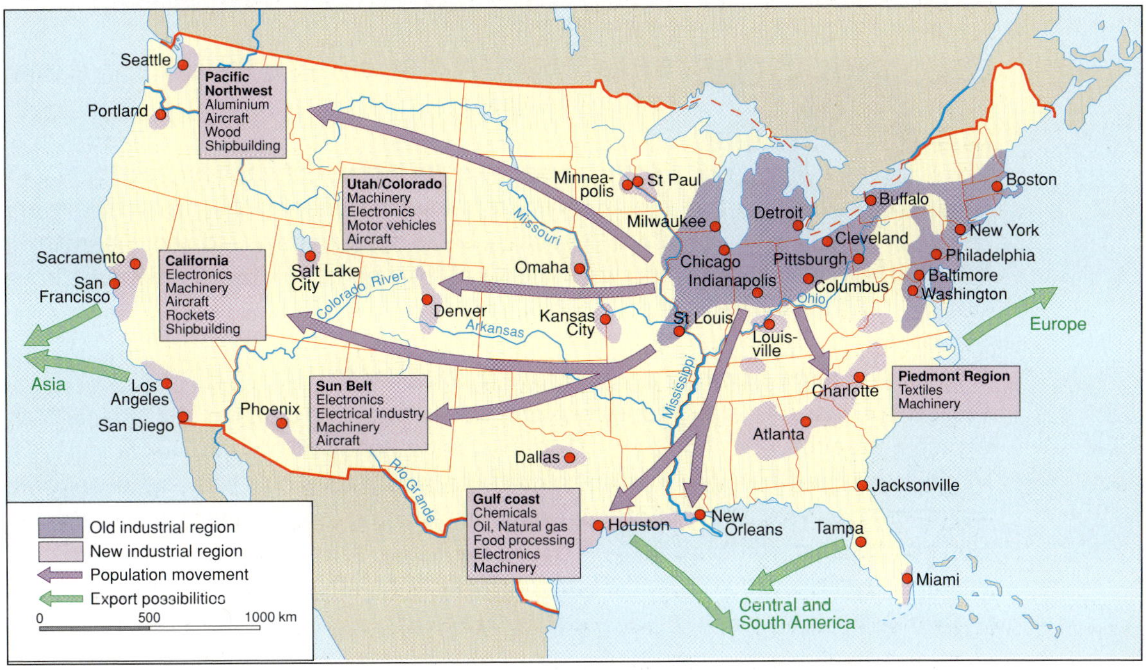

GEOGRAPHY TASKS

1 List specific reasons why the Northeast of the USA became known as the Manufacturing Belt.

2 Why is this area now often called the Rust Belt?

3 Describe the distribution of the industries within the Manufacturing Belt and give reasons for this pattern.

4 Use your atlas and other sources of information to collect facts about the industrial regions of Germany. Find similarities and differences between the Manufacturing Belt in the USA and the main industrial regions in Germany.

5 Using the map in Figure 14, explain where the new industrial regions are located. What are the main types of industry in the South and the West? Explain why these industries are attracted to the Sun Belt.

6 Use Figure 15 to calculate the total number of people living in the Snow Belt and the Sun Belt in 1950 and 2000. Draw two lines on a graph to show the change in population in each region, and describe what the graph shows. Give reasons for the different rates of population growth in the two regions.

Figure 15 **Population change in the Snow Belt and the Sun Belt (in millions)**

Snow Belt	1950	2000
Massachusetts	4.7	6.4
Rhode Island	0.8	1.1
Connecticut	2.0	3.4
New York	14.8	19.0
New Jersey	4.8	8.4
Pennsylvania	10.5	12.3
Ohio	7.9	11.4
Indiana	3.9	6.1
Illinois	8.7	12.4

Sun Belt	1950	2000
Florida	2.8	16.0
Mississippi	2.2	2.8
Texas	7.7	20.9
Arizona	0.8	5.1
Nevada	0.2	2.0
California	10.6	34.0
Wyoming	0.3	0.5
Colorado	1.3	4.3

THE VALLEY OF THE SUN – PHOENIX, ARIZONA

Rachel Kibler-Melby, 14, moved to Phoenix from a small town in Minnesota.

Rachel, what was the biggest change for you?
The weather, of course! In Phoenix it's sunny almost every day. I like sports, but in the summer it's too hot for anything but swimming. Of all my friends, only two of them don't have a pool in their backyard.

Why do you think so many people move to Phoenix?
Well, in our case it was my mom's job. She was called to be a pastor at a big Lutheran church with about 2000 people every Sunday. I have friends from North Dakota and West Virginia who came here because of their parents' jobs. And people come because of the warm weather. We have lots of "snowbirds" in Arizona, mostly older people from the North or the East who only live here during the winter. And there are lots of retirement communities around here, too, like Sun City. They're small towns where you can only live if you're over 55.

What do kids in Phoenix like to do for fun?
In the summer, when we're not in a pool, we try to find someplace cool. My friends and I like to go to a big shopping mall or to a movie where everything is air-conditioned. In the winter, there's snow in the mountains only about two hours away, so sometimes we go sledding or skiing. In our family, we like to hike too, especially when the cactus plants are in full bloom. There are also lots of interesting old Indian ruins to explore.

Are there any disadvantages of living in Phoenix?
We live in a nice neighbourhood on the north side of town near a famous golf course, but there are places in Phoenix my parents won't let me go. Especially on the south side, there's a lot of crime and problems with gangs and guns. At our church we collect food and blankets for homeless people and help at soup kitchens.

Sun City, near Phoenix

- Phoenix is one of America's fastest growing cities. The population in 1940: 30,000; in 1960: 439,000; in 1980: 790,000; in 1995: 1,083,000. But in 1858, before Phoenix was founded, an American government official had said, "The region is altogether valueless. After entering it, there is nothing to do but leave".
- Water and air-conditioning made the difference. A system of canals and reservoirs brings water to the desert city. Some of the canal systems are similar to those used by Native Americans over 500 years ago. Air-conditioning makes it possible to live and work in a climate with temperatures over 40°C every day in the summer.
- During World War II, the clear, dry air made the area an ideal place for military pilots to train. Lufthansa also has a large pilot-training centre in the Phoenix area.
- Over half of Arizona's manufacturing jobs are in high-tech industries, compared to 20 per cent in the rest of the USA. Many businesses are attracted to Phoenix because land is relatively cheap and taxes and labour costs are lower than in many other cities.

Tuscaloosa County Industrial Development Authority

File Edit View Go Help

Back Forward Reload Start Search Guide Print Security Stop

Bookmarks Address: http://www.tcida.com/

TUSCALOOSA
An International Attraction

Next Stop | Last Stop | Tour Itinerary | Site Index | Contact TCIDA

Welcome to Tuscaloosa County, Alabama

This virtual tour will demonstrate why Tuscaloosa represents the "best" of so many things, including "the best" location for manufacturing and service industries looking to succeed in today's global market. Most recently, the Tuscaloosa area was recognized by *Kiplingers' Personal Finance Magazine* as one of the nation's 10 best "entrepreneurial hot spots", and by *Money* magazine in its annual list of the "most livable" cities in the United States. And with the selection of Tuscaloosa County by Mercedes-Benz as the site to produce its new M-Class, All-Activity Vehicle, the best just keeps getting better.

Mercedes selected Tuscaloosa because of its "quality work force, strong transportation network, the University of Alabama, business vitality and quality of life". By the end of this century, Mercedes expects to employ 1500 people.

Location! Location! Location!

Tuscaloosa County is strategically located close to key, major markets in the Southeast. The area is 50 miles southwest of Birmingham and 95 miles northwest of Montgomery. Atlanta, Memphis and Nashville are within a 200-mile radius. Mobile, New Orleans and the Gulf Coast of Alabama are just a few hours to the south. In West Alabama, where the Appalachian Mountains meet the coastal plain of Alabama and the Black Warrior River flows through, you will find Tuscaloosa. Tuscaloosa's geography is a physically diverse area, with hills and mountains to the north, and flat, fertile land to the south. Ever present are shimmering lakes and rivers, and an abundance of rich natural resources. The area enjoys a moderate climate with an average temperature of 63.4°F.

An International Influence

The slogan "Tuscaloosa ... An International Attraction" proves more true with each day ... International Supplementary Schools recognize the concern of parents that their children be able to reenter and adapt to their native country's education system after attending local Alabama schools. Programs currently exist for students native to Germany and Japan ... And Tuscaloosa continues to encourage cultural growth of all its citizens through festivals recognizing our international relationships. International CityFest represents Tuscaloosa's relationship with Mercedes-Benz while the Sakura Festival celebrates Tuscaloosa's friendly ties with Japan and its sister city – Narashino City.

Document: done

GEOGRAPHY PROJECTS

1. **Investigate why foreign companies are building factories in the USA. Look at factors such as location, climate, incentives for new businesses, growth industries, changes in population, and problems caused by rapid growth.**
2. **Las Vegas is another fast-growing US city. Find out using the Internet and other sources why this city is so attractive to both tourists and businesses.**

Atlanta, Georgia

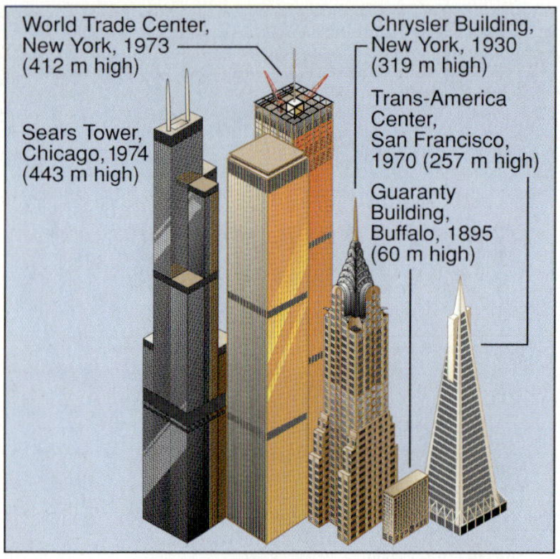

Figure 16 **The development of skyscrapers in the USA**

SKYSCRAPERS AND SUBURBS

Compared to old European cities that grew up slowly around a marketplace, most US cities are young. Instead of medieval churches and narrow streets, the centre of a typical American city is a group of skyscrapers. While New York's impressive skyline is probably seen on more postcards than any other city's, skyscrapers were actually first built in Chicago and the tallest building in the USA, the Sears Tower, is in Chicago.

Another thing that is typical for US cities is their checkerboard pattern of endlessly long and straight streets. This can make the cities seem boring, but it also makes it relatively easy to find a place if you know the address. The house numbering system in a city starts from a central point, and in many cities the streets have numbers instead of names (such as the famous Fifth Avenue in New York) which makes it even easier to see where you are. Often there is also a system where "streets" go from east to west and "avenues" go from north to south.

When immigrants from other countries come to America, they often settle in neighbourhoods with other people from the same home country. "Chinatown", "Little Italy" or "Little Tokyo" are typical names for these neighbourhoods. Thousands of people are still immigrating to America every year and you can find areas in cities where most of the shop signs are in Vietnamese or Korean or Spanish.

While it is true that over 75 per cent of all Americans live in cities, it is also true that about three-quarters of all Americans live in a house of their own. This means that the cities are very spread out with street after street of single-family houses. It also means that shops, jobs, and schools are seldom within walking distance, because most areas were built up after the invention of cars. In general, most American cities do not have a good public

Residential area in Minneapolis

transport system because of the great distances and because people are used to the independence of their own cars. Of course this also leads to traffic and air pollution problems. Los Angeles is particularly famous for its crowded freeways and smoggy air. Strict laws for smog control devices on cars have improved the air pollution situation, and the city is working on a city-wide train system, but it is still not unusual for people to spend at least three or four hours a day in their cars, moving slowly along the crowded freeways which are sometimes called "the world's longest parking lot".

Because so many people want to live in a single house with a garden, the built-up areas around the cities, called the suburbs, have spread out further and further creating huge areas of nothing but streets and houses and shopping centres. It is often misleading to talk about the population of a city because the area around it, called the metropolitan area, is often several times larger. Only about 700,000 people live directly in San Francisco, for example, but there are over six million in the metropolitan area. In some places, several cities and metropolitan areas run together to create a huge built-up area called a megalopolis. The largest is "Bos-Wash", between Boston and Washington, DC, covering a distance of around 600 kilometres, where over 50 million people live on less than 2 per cent of the total land area of the USA.

Figure 17 **A megalopolis along the East Coast**

Figure 18 **Cities with over one million inhabitants (2000)**

	New York	Los Angeles	Chicago	Houston	Philadelphia	Phoenix	San Diego	Dallas	San Antonio
City	8,008,300	3,694,800	2,896,000	1,953,600	1,517,500	1,321,000	1,223,400	1,188,600	1,144,600
Metropolitan area	21,199,900	16,373,600	9,157,500	4,669,600	6,188,500	3,251,900	2,813,800	5,221,800	1,592,400

Source: Spiegel Almanach, 2002; Time Almanach, 2002

GEOGRAPHY TASKS

1 **Compare the population of each city listed in Figure 18 with that of its metropolitan area. Explain why the numbers are so different.**

2 **Find the large cities and metropolitan areas on a map. Describe the locations of these cities and list their possible attractions.**

3 **What are some advantages and disadvantages of living in a single-family house? What are some of the consequences for society when most people live in a house of their own? In what kind of house would you prefer to live and why?**

4 **Most US cities were built more recently than German cities. What difference has this made in their layout and planning?**

5 **Suggest reasons why immigrants often settle in parts of a city which are dominated by their ethnic group. Consider large German cities like Berlin and Cologne.**

......Russia

The **Russian Federation** (in Russian: *Rossiyskaya Federatsiya*) is the official name of the country usually called simply **Russia**. It is the largest country in the world, covering one-third of Asia and two-fifths of Europe.

The **Commonwealth of Independent States** (**CIS**) has existed since 1991. It is a group of independent countries with a central authority a bit similar to the idea of the European Union. Russia is by far the largest and most powerful of the 12 countries in the CIS. The purpose of the group is to co-ordinate the economic and foreign policies of the countries as well as to work together on environmental and crime issues. From the beginning the group has been unstable with many economic and ethnic conflicts among the member countries.

The **Soviet Union**, also called the **Union of Soviet Socialist Republics** (**USSR**), existed from 1922–1991. It was one huge country which included the present CIS nations plus Estonia, Latvia and Lithuania, which are now independent. The Soviet Union was the first and largest communist country in the world. The state owned all the land and factories and controlled the economy through a central plan instead of a free-market economy. The state organised schools, jobs and health care for all the people, but there was very little political freedom. Many people who disagreed with the government were sent to prison. Especially in the 1930s millions of political prisoners were murdered.

Before the Russian Revolution in 1917, Russia was ruled by czars. A relatively small country in the 1300s, Russia became larger and larger by conquering more and more land until it controlled one-sixth of all the land in the world. Throughout Russian history there were always very big contrasts between the riches of the czars and the poverty of the masses of people.

A country of many different ethnic groups

Today's Russian Federation is one of the most multicultural countries in the world with over 100 different ethnic groups and languages. Russian is the official language and is taught in all the schools, but almost 20 per cent of the population is not Russian. During the time of the Soviet Union many people were forced to move to different parts of the country. Only Russians were given positions of leadership in politics and industry, even in places where the Russians were a minority. This helped keep the country stable, but it is one cause of many ethnic conflicts today.

Figure 1 The Commonwealth of Independent States (CIS)

Figure 2 Members of the CIS (2002)

Country	Capital	Area (1000 sq km)	Population (millions)	% of population who are Russian
Armenia	Yerevan	30	3.3	2.0
Azerbaijan	Baku	87	7.8	2,5
Belarus	Minsk	207	10.4	13.2
Georgia	T'bilisi	70	5.0	6.3
Kazakhstan	Almaty	2717	16.7	34.7
Kyrgyzstan	Bishnek	198	4.8	18.0
Moldova	Chisinau	33	4.4	13.0
Russia	Moscow	17,075	145.5	81.5
Tajikistan	Dushanbe	143	6.6	3.5
Turkmenistan	Ashgabat	488	4.6	6.7
Ukraine	Kyyiv	603	48.8	22.0
Uzbekistan	Toshkent	447	25.2	5.5

Figure 3 The ten largest countries of the world (2001)

Country	Area (million sq km)	Population (millions)
1 Russia	17.1	145.5
2 Canada	10.0	31.6
3 China	9.6	1273.0
4 USA	9.6	278.0
5 Brazil	8.5	174.5
6 Australia	7.7	19.4
(Europe, without Russia*)	5.7	581.0
7 India	3.3	1030.0
8 Argentina	2.8	37.4
9 Kazakhstan	2.7	16.7
10 Algeria	2.4	31.7
Total	**73.7**	**1766.1**

GEOGRAPHY TASKS

1 **Calculate the population density in each country of the CIS. Then, working in pairs, draw a density-shaded map to show population density, and a second density-shaded map to show the percentage of the population of each state which is Russian. You can use the map above as a basis for your drawings. Describe the patterns on the two maps, and note any similarities between them.**

2 **Compare the CIS figures with these figures for Germany:**

Area: 357,022 sq km, Population: 82,183,000. Which CIS countries are most/least similar to Germany in area, population, and population density?

3 **Use information from Figure 3 to describe Russia's ranking in area and population in a world-wide context. What reasons could there be for the difference in ranking in area and population?**

ACROSS RUSSIA WITH THE TRANS-SIBERIAN RAILWAY

Because Russia is such a big country, transportation between one end and the other is an important problem. In the 1890s the Russians began building the longest railway in the world to connect Moscow in the west to the Pacific Ocean in the east. Before the railway was built, the trip took about a year along very rough roads. One reason for building the "Trans-Sib" was military: to transport troops to the East in order to protect the country against Japan and China. Another reason was to transport food from the agricultural areas of south-western Siberia to the people in the European part of Russia. Many cities and industries were built up along the route of the Trans-Siberian Railway.

Monday Our train left Moscow shortly after 2 pm. At first we only saw factories and blocks of flats, then countless little summer-houses. For hours and hours we travelled across the Russian plains through mixed forests, farmland and meadows.
Tuesday Shortly before Yekaterinburg we crossed the Ural Mountains, which here are rolling hills and not mountains. This area has mines, steelworks and other factories. We are leaving Europe now and going into Asia – to be exact, Siberia.
Wednesday This is the busiest part of the railway line, between Omsk and Novosibirsk. The scenery here in the lowlands of western Siberia: flat land, rivers, lakes, birch forests, wheat fields.
Thursday After Tajšet our route went along the southern side of the central Siberian mountains passing rounded mountain tops, coniferous forests, gentle valleys and wooden rafts on the rivers. After passing through Irkutsk we saw Lake Baikal, the "Siberian Ocean". At 1620 meters it is the deepest lake on earth and contains more water than the Baltic Sea.
Friday Today we crossed the Yablonovy mountain range, an altitude of 1000 m. These mountains form the watershed between the rivers that flow into the Arctic Ocean and those that flow into the Pacific Ocean. We have left Siberia and are in Russia's Far East.
Saturday We travelled through the Amur lowlands, past endless mixed forests with oak, lime and nut trees. It reminded us a little bit of the area west of the Urals.
Sunday People call the area around the city of Khabarovsk the "golden ground", not only because of bits of gold in the river sediments. The land is fertile and there are forests, mineral resources and many fish in the rivers.
Monday After seven days and nights we finally reached the Pacific harbour of Vladivostok. A plaque at the train station reads, "Here Ends the Great Trans-Siberian Railway. Distance from Moscow: 9288 km".

GEOGRAPHY TASKS

1 Draw an outline map of Russia and mark the route of the Trans-Siberian Railway using information from the text. Put on your map the rivers, mountain ranges and important towns that you would go through if you travelled from Moscow to Vladivostok.

2 Use your atlas to measure the distance between the eastern and western borders of the CIS.

3 How many time zones does the Russian Federation have? What does this mean for the daily life in a country?

4 Name the vegetation zones that you would cross if you went from the Caspian Sea to the Kara Sea.

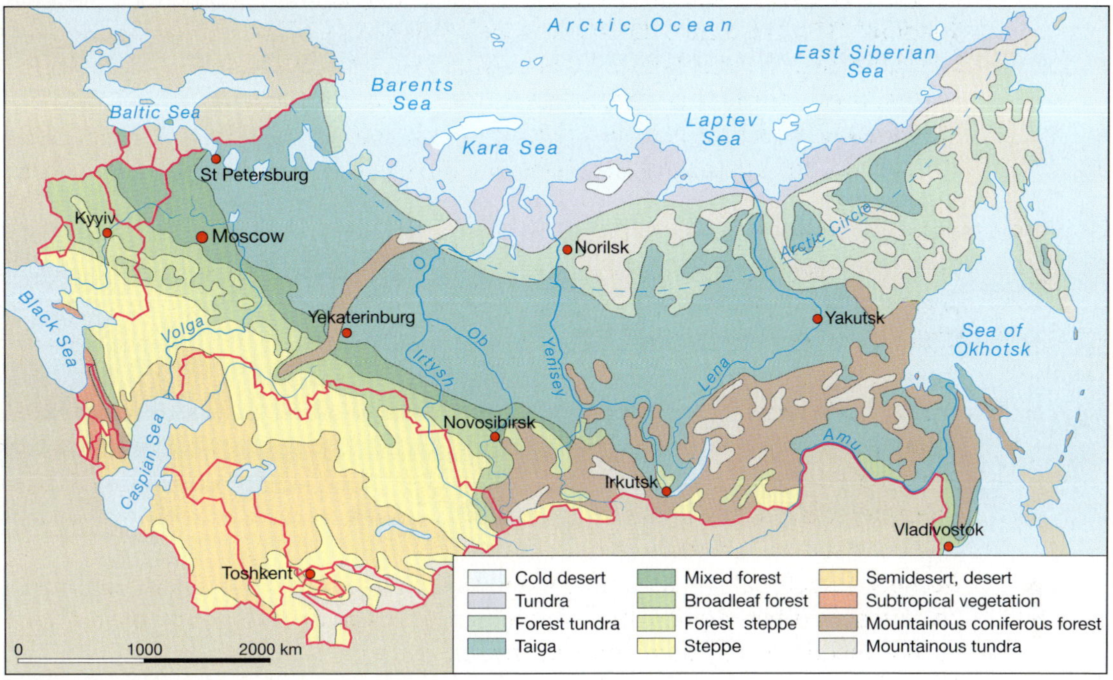

Figure 4 **Vegetation zones**

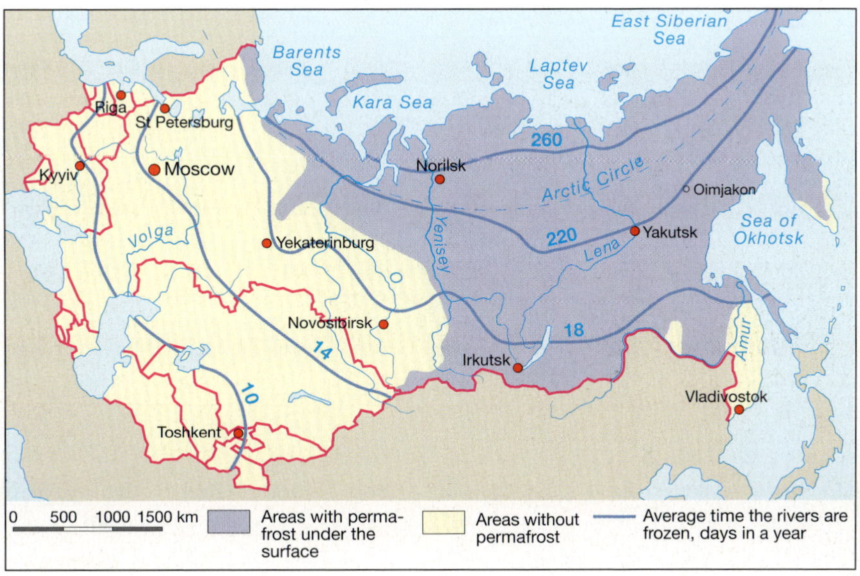

Figure 5 **Distribution of permafrost and average number of days the rivers are frozen**

GEOGRAPHY TASKS (CONT.)

5 The photos of four natural regions in Russia are shown opposite in the correct sequence from north to south, but the texts and climate graphs are mixed up. Find the text and graph that goes with each photo and identify the region using one of the following terms:

Steppe, Semidesert/Desert, Tundra, Taiga.

A *The growing period for plants here is 120 days. Winter temperatures are under –20°C. Spruce trees that grow 30 m tall are cut down for sawmills. Wood from Boreal coniferous forests is very solid and has fine, narrow annual rings.*

B *Precipitation is very low here all year. In summer the temperature reaches 25°C. There is very little vegetation. Agriculture is only possible with irrigation.*

C *The temperature range here is over 40°C. Herding reindeer is the only kind of agriculture possible. Mosses and lichens grow here as well as birch trees the size of walking sticks.*

D *Plants here have a growing period of 150 days. Thanks to the fertile black soil, large areas of this region can be used for wheat monoculture.*

1	Jan	Feb	Mar	Apr	May	Jun	Jul	Aug	Sep	Oct	Nov	Dec	Year
°C	–18.1	–17.6	–11.1	0.5	11.2	17.1	19.5	16.5	10.8	1.8	–8.5	–16.3	0.4
mm	8	6	9	18	30	53	72	46	33	23	15	12	325

2	Jan	Feb	Mar	Apr	May	Jun	Jul	Aug	Sep	Oct	Nov	Dec	Year
°C	–22.2	–19.3	–12.8	–4.4	3.6	12.6	16.8	13.9	7.4	–1.7	–13.3	–20.2	–3.3
mm	24	19	23	30	48	55	68	57	60	49	30	29	492

3	Jan	Feb	Mar	Apr	May	Jun	Jul	Aug	Sep	Oct	Nov	Dec	Year
°C	–6.9	–5.8	0.2	9.3	17.7	22.5	25.1	23.3	17.1	9.9	2.3	–3.5	9.3
mm	11	11	9	15	21	21	17	13	17	11	12	17	175

4	Jan	Feb	Mar	Apr	May	Jun	Jul	Aug	Sep	Oct	Nov	Dec	Year
°C	–30	–26	–23	–16	–6	–4	12	10	3	–8	–22	–27	–11
mm	12	11	9	10	12	29	32	49	33	28	18	13	267

Changes in Agriculture

Russia is a huge country, but almost 90 per cent of its land cannot be used for agriculture. It is either too cold or too dry or too mountainous. The main farming area is within the so-called fertile triangle: from St Petersburg to the Black Sea to southern Siberia. But even in the best farming areas, some years there is either too much or not enough rain.

Besides climate problems, Russian agriculture has many structural problems as the country tries to change from a centrally planned communist economy to a free-market economy. One of the main ideas of the Russian Revolution of 1917 was that all land should belong to the state and not to private individuals. During the time of the Soviet Union, all the farms were organised either as collective farms *(kolkhozes)* or as state farms *(sovkhozes)*. The farms were not as productive as those in western Europe and North America, partly because of a lower level of technology and partly because of climate conditions.

With the change to a market economy, the ownership of the collective farms was turned over to the members of the collective as a group. Individuals were also allowed to become private owners. The new system has many serious new problems. Russian agricultural production declined by about ten percent every year in the 1990s. For some products, production levels were the lowest in 30 or 40 years. With fewer price controls, the cost of fertiliser, seed, and machinery has gone up drastically, but the state can no longer afford to pay the farmers subsidies or give them credit. The prices of farm products on the free market are so high that most people cannot afford to buy them: over half of all Russians are now growing their own fruit and vegetables in garden plots. This means that the domestic market for farmers' products is getting even smaller. But the farmers cannot sell their products on foreign markets, because countries with a higher level of productivity can grow things much cheaper. In the old days, the quality of the products was not so important because the farm workers always received the same amount of money anyway. There was no motivation to improve the products or the productivity. Now Russian farmers have to change not only their methods of production but also their way of thinking. The farmers who want to change need a lot of money to invest in new improvements, but getting enough credit is the biggest problem of the new Russian farmers.

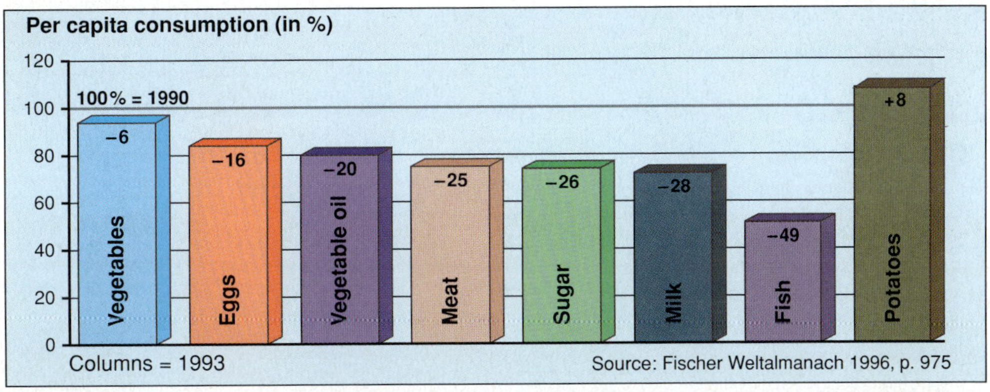

Figure 6 **Decline in per capita consumption of food 1990–1993**

Crop	Russia			Germany
	Highest yield (by economic region)	Lowest yield (by economic region)	Average yield	Average yield
Grain (dt/ha)	22 (Northern Caucasus)	7 (Far East)	11	58
Potatoes (dt/ha)	184 (European North)	66 (Northern Caucasus)	117	607
Vegetables (dt/ha)	203 (European North)	83 (Northern Caucasus)	142	272
Sugar beet (dt/ha)	216 (Northern Caucasus)	132 (Ural)	176	484

Figure 7 **Yields of some agricultural products in the Russian Federation and Germany**

In Russia, what Valentin Dimitriyov does is so unusual that the Russian language had to import an English word for it. He is called a fermer, *or farmer, someone who owns and works on their own private farm.*

Valentin, 52, and his family have 60 hectares of land. Their new two-storey house in Gordina, a small town 150 km west of Moscow, is almost finished. The Dimitriyovs have five old tractors, three trucks and two cars. They buy scrapped old engines to use for spare parts. They grow potatoes and have 300 pigs and 18 cows. That should be enough to make a good income for the five-member family. But there are some problems:

Valentin always worries when the electric heating system that keeps his young pigs at the right temperature breaks down. The state AgroService often sends out its technicians so late that the pigs do not survive. Another problem is that there is so much work that he, his wife Katya and their oldest daughter cannot do it all themselves, but it is very hard to find good, reliable workers. Valentin also worries about his buyer, the state-owned packing plants. They take advantage of their monopoly position and set such low prices that they do not even pay his expenses. He has not been able to find private buyers for his products.

With the help of a newly-founded farmers' association, they hope to break the state's monopoly. Valentin will take out an inexpensive loan from the association and hopes to open his own business in the next town. There will be a store where over a hundred fermers *can sell their products. If this plan works out, the Dimitriyovs will be among the 20 per cent of the farming operations around Moscow that have survived the change to a market economy.*

	Russia	**Germany**
Land used for agriculture (in million hectares)	210	20
Percentage of total land area used for agriculture	12	55
Percentage of people working in agriculture	**12.9 (1990)** **15.0 (1995)**	**3 (1994)**

Figure 8 **Agriculture in Russia and Germany**

Kind of operation	**Percentage of total agricultural land**	**Number of operations**	**Percentage of total agricultural production**
Kolkhoz	14.9	5800*	
Sovkhoz	8.3	3500*	55
Joint stock companies, cooperatives and other reorganized former kolkhozes and sovkhozes	50.3	17,700	
Private operations of kolkhoz and sovkhoz members	3.7	38,700	43
Single farmer operations	5.6	280,100	2
Other non-agricultural uses	17.2		

*estimate

Figure 9 **Ownership and structural forms of Russian agriculture (1995)**

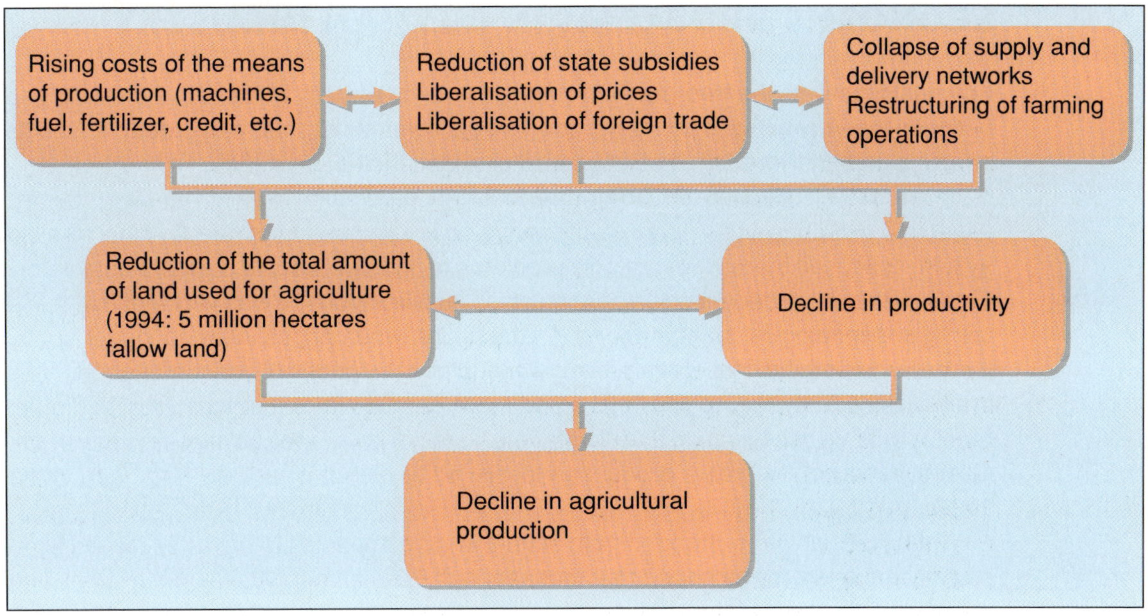

Figure 10 **Causes of the crisis in Russian agriculture**

PRACTISE YOUR ENGLISH

You can use the following expressions to show the relationships in the flow chart:

The rising production costs and / as well as / together with the reduction in state subsidies caused / led to / influenced the decline in productivity and ...

State subsidies were ... ... reduced.
Farming operations were ... ... restructured.
Prices were ... ... liberalised.
One thing that led to a decline in productivity was ...
Another factor was ...
As a consequence/result ... This in turn led to / caused ...

GEOGRAPHY TASKS

1 **List four problems faced by the Dimitriyov family since the change to a market economy.**

2 **Compare the data on Russian and German agriculture. What are some reasons for the differences in the percentage of land used for agriculture and the percentage of people working in agriculture? What reasons could there be for the increase in the number of people working in agriculture in Russia?**

3 **Describe the main organisational change that has taken place in the structure of Russian agriculture.**

4 **Using information from the flow chart in Figure 10, write a short text about problems in Russian agriculture.**

ECONOMIC PROBLEMS IN RUSSIAN INDUSTRY

For most Russians the free-market system is a hard new way of life with a growing gap between the very rich and the very poor. The old communist state had greatly limited its people's personal freedom, but it had also provided their basic needs.

In the state-controlled economy the focus of production was on heavy industry, especially weapons and military equipment, and not on consumer goods. When the old system collapsed there were not only problems to find new trading partners outside the Eastern Bloc, but it became clear that much of Russia's industry was very inefficient and with low technological standards. As in agriculture, when prices were no longer set by the government, the cost of production went up and the prices of consumer goods rose dramatically. At the same time more and more people either lost their jobs or did not receive their wages because their firms simply did not have enough money to pay them. Over a period of five years production levels fell to half of what they had been in the 1980s.

When Russian firms changed from state-owned companies to privately-owned joint-stock companies, it was hoped that thousands of Russian people would buy stock and become a strong new middle-class. So far this has not happened. Instead, a very small group of Russians has managed to get very rich through business deals, often through connections to politicians and special tax privileges. The state suffers from not enough tax money coming in to pay government workers, teachers, soldiers, police officers, as well as pensions, health care, roads, social services, and many other things. Industries are either not making any money, or if they are, they find ways to avoid paying taxes. Russia also has a huge and growing foreign debt.

One way out of the economic crisis is for foreign companies to invest in Russia and build factories and stores there. Many new businesses are organized as joint ventures between a foreign company and a Russian company with the hope that both companies and both countries will benefit from the system.

Kirov Plant Near Bankruptcy?

In St Petersburg, the former centre of the Soviet Union's weapons industry, many of the most important industrial firms are close to financial disaster. In the Kirov plant, once a leader of Russian industry, only 14,000 employees are left from 43,000 five years ago. The company stopped making tanks two years ago. They used to make 50 "Kirovets" tractors a day, then only five, and now they cannot even sell that many. Conflicts between Russia and Kazakhstan, competition from Canadian tractor firms, the high costs of production, and the financial problems in Russian agriculture are some reasons for the crisis. Additional problems are out-of-date technology and management's inability to adapt to the new conditions.

(adapted from FAZ, *July 6, 1994)*

	1990	1991	1992	1993	1994	1995
Thermal energy	102.0	100.3	95.5	95.0	91.0	97.7
Fossil fuel industry	97.0	94.0	93.0	88.0	90.0	99.2
Iron works	98.0	93.0	84.0	83.0	83.0	110.0
Non-ferrous metalworks	98.0	91.0	75.0	86.0	91.0	103.0
Chemical/Petrochemical industry	98.0	94.0	78.0	79.0	76.0	108.0
Machine industry	101.0	90.0	85.0	84.0	69.0	91.0
Wood industry	99.0	91.0	85.0	81.0	70.0	99.3
Light industry	99.9	91.0	70.0	77.0	54.0	70.0
Food production/processing	100.4	91.0	84.0	91.0	83.0	92.0
Total	**99.9**	**92.0**	**82.0**	**86.0**	**79.0**	**97.0**

Figure 11 **Industrial production by branch, 1990–1995 (in % of production for previous year)**

Gas from Russia

Russia has more natural gas than any other country in the world. Much of it is sold to Europe. To get from Siberia to Germany, the gas travels through a pipeline that is almost 6000 km long.

Country	Billion m^3
Germany	27.2
France	14.3
Italy	13.7
Czech Republic and Republic of Slovakia	3.5
Romania	7.2
Hungary	5.0
Austria	5.0

Figure 12 **The leading importers of Russian natural gas**

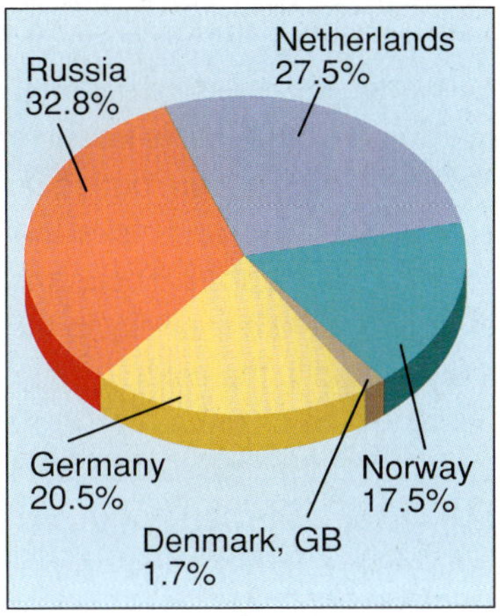

Figure 13 **Where gas in Germany comes from**

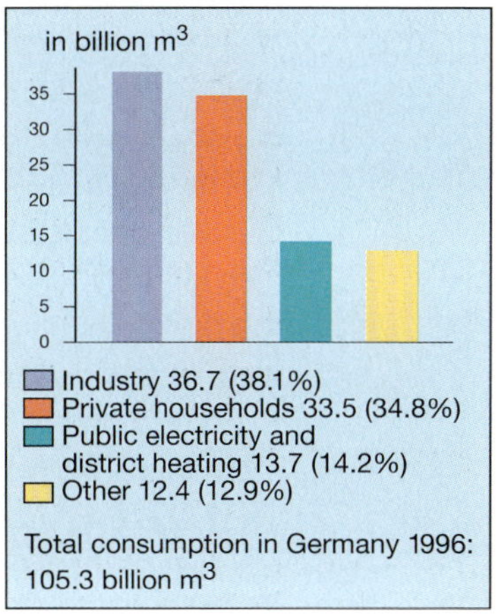

Figure 14 **How gas is used in Germany**

Juri Ryshkov, 28, is from Ufa in the European part of Russia. He works in a gas-producing area of northern Siberia. His wife and son live 2000 km away. Juri sees them every two weeks. Building and maintaining permanent housing in the tundra would be too expensive, so the workers are flown back and forth to their homes. During his two-week working shift, Juri and his colleagues live in simple barracks and "living tanks". Every day they drive in tracked vehicles across the tundra to the gas fields and pumping station. "When we put the gas under pressure to get it ready for transport, it gets very warm. We have to keep it cooled to below zero. Otherwise the permafrost ground would thaw and everything would sink. Sometimes there are breaks in the pipes that aren't discovered and repaired until some days later, so there is always the chance of an accident."

Russian gas pipeline explodes

A natural gas pipeline exploded yesterday in the northern Russian industrial region of Uchta, an area which has seen many environmental disasters. For several hours flames from the escaping gas shot high into the air and set fire to large areas of forests. The cause of the accident, which so far has not taken any human lives, is the old and badly serviced equipment. A commission from the Russian parliament estimates that each year about five million tons of natural gas and mineral oil are lost through hundreds of accidents. (adapted from the Stuttgarter Zeitung, *April 28, 1995)*

GEOGRAPHY TASKS

1 **Find three problems which agriculture and industry in Russia have in common.**

2 **Figure 11 shows a five-year trend in production. Divide the industries into three groups showing: (1) an overall increase in production, (2) a decline in production and (3) little or no change. Suggest reasons for the declines.**

3 **Find out as many uses as you can for natural gas.**

NORILSK – AN INDUSTRIAL "ISLAND" IN THE SIBERIAN TUNDRA

Why would anyone build a city of over 200,000 people in a place north of the Arctic Circle where the winter temperature is often –50°C and there is snow 250 days a year, where no trees and no farm crops can grow because of the permafrost, where the ground becomes a swamp in summer with massive swarms of mosquitoes, and where the next big city is 2000 km away down a river that is frozen from October to March?

The only answer is the treasures that are buried under the frozen ground. In the 1930s Russian geologists discovered copper and nickel there as well as coal and iron ore. They also found platinum, palladium and cobalt. Thousands of political prisoners were sent up the river to work in mines and build a new city in the Siberian tundra. Living conditions were terrible and thousands died in the cold.

Later, workers also went to Norilsk because the state promised to pay them four times more than they would receive on the "mainland", as Norilsk people call the rest of Russia. They are on an isolated "island" in the north, with no roads at all connecting their city with the rest of the world, and where the nearest railway station is 1000 km away. But the "Kombinat", as the industrial complex was called, took care of the workers and their families with schools, hospitals, and canteens. They even gave some of them a two-week holiday at the Black Sea every year.

Those days are over. As in the rest of Russia, industry in Norilsk is facing many problems. Its technology is old and inefficient, it cannot pay all the workers, several thousand have already lost their jobs, and the city cannot afford the social services. Prices have risen so high that people who had hoped to leave Norilsk when they had saved enough money have lost that dream. They are glad if they still have a job at all.

Another big problem is that Norislk is an ecological disaster area where you can hardly see the smokestacks through the yellowish-grey haze of sulphur dioxide. As one unemployed worker said, "We've given them our lungs for their metal".

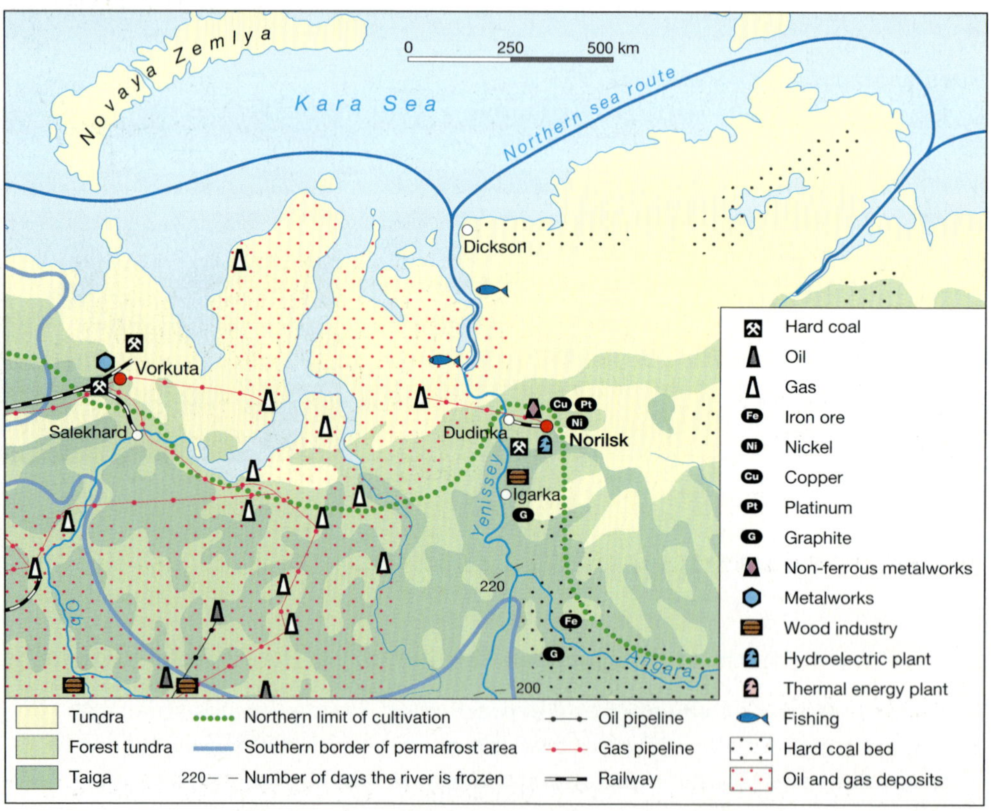

Figure 15 **Northern Yenisey region with natural resources**

Street scene in Norilsk

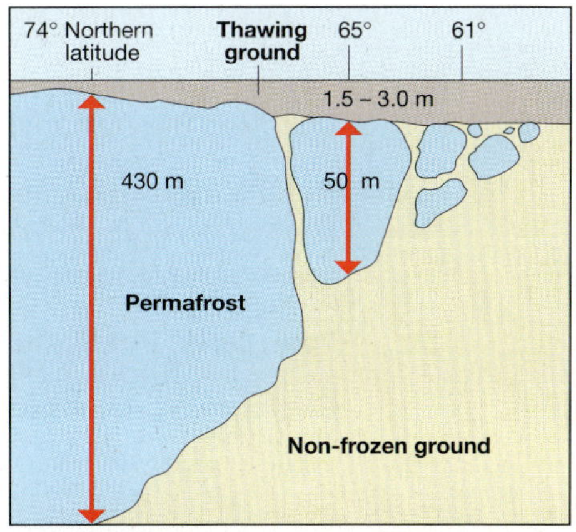

Figure 16 The ground under Norilsk

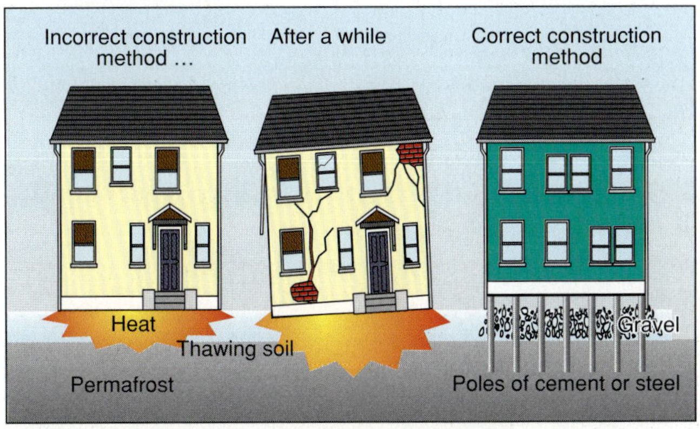

Figure 17 Construction in permafrost regions

Houses built on poles

PRACTISE YOUR ENGLISH

Match the *heads* and *tails*. Then put the sentences in the order that describes how houses are built on permafrost. Explain the method using connecting words such as: first of all, then / after that, etc.

Heads	**Tails**
Water …	… it is as strong as cement.
If the houses were not supported by poles …	… are drilled into the icy ground.
Holes up to 12 metres deep …	… is poured into the holes around the poles.
When the water freezes …	… are put into the ground.
Poles made of iron and cement …	… they would crack when the ground melted.

GEOGRAPHY TASKS

1 **How can the mineral resources of the Norilsk region be exported to the rest of Russia and the world? What difficulties does this transport present?**

2 List the natural resources and types of industry found in the Northern Yenisey region.

3 Describe the climate and vegetation of the area around Norilsk.

4 Put the following data into a line graph showing the population growth of Norilsk. Explain how Norilsk could grow so enormously in spite of its unfavourable location.

Year: 1939, Population: 14,000; 1959 – 108,000; 1979 – 205,000; 1989 – 220,000.

5 Explain why the airport in Norilsk needs a cooling system under its runway.

MOSCOW

In Moscow
... 20% of all Russian businesses have their headquarters.
... 50% of all joint ventures with foreign investors are located.
... 75 of the 100 large banks have their headquarters.
... 25% of Russia's total business is transacted.
... about one-third of Russia's total tax income is earned.
... rent for shops in top locations is over $8000 per square metre per year.
... all important economic and political decisions are made.

In the days of the Soviet Union, if you saw a car on the grey streets of Moscow you could be quite sure that there was a government official inside. It was rare for a private citizen to own a car. At the end of the communist time in 1991, there were only 700,000 cars in Moscow, a city with over nine million people, three times more than Berlin. By 1997 the number of cars had jumped to over three million. Traffic jams have become a part of modern Moscow, along with fast-food restaurants, luxurious new banks and office buildings, designer boutiques, fitness centres, 24-hour supermarkets, and wild night clubs. Moscow changed more in half a decade than in the previous half century.

Moscow is the centre of industry, trade, and political power in Russia, as it has been for hundreds of years. It is also the centre of the new Russia with great contrasts between winners and losers in the new economic system. Some people are making lots of money in the new market economy, both legally and illegally. But there are also many people who have lost their jobs, or if they still have a job, they do not get paid regularly. Prices in the shops are so high that many people have become pedlars and try to sell whatever they can on the streets or on trains.

But in spite of many problems with crime, unemployment, air pollution and a lack of social services, people are still attracted to Moscow. Over 1800 trains come and go from the Russian capital every day, and the city has five international airports. Besides the nine million who live there, another two or three million come into the city every day to work or look for work. Hundreds of foreign business people who are trying to set up new businesses in Russia have their offices in Moscow. There are also always lots of tourists in Moscow.

One of the most famous attractions is the Kremlin and Red Square, the heart of Moscow. The Kremlin, which is an area with old palaces, cathedrals, and official buildings, was first built as a fort in the 12th century. Red Square, the site of many political demonstrations, was once a marketplace. The city of Moscow grew up around these landmarks. When you look at a map of Moscow, you can see some rings around the city that were once city walls and now are roads. The most recent ring is a four-lane motorway that was built in the 1960s and is one of Russia's few motorways. The city's population continues to grow and spread outward. Moscow together with the built-up area around it now covers over 1000 square kilometres.

In the days of the Soviet Union no one could own their own flat or house. That is changing quickly and there is a building boom all around Moscow. People who can afford it are buying prefabricated houses or modern flats, or are renovating old buildings. Many people are making little dachas, summer houses in the countryside, into places to live all the time. Compared to Western Europe or North America, the number of people who own their own house or flat is still very low, under ten per cent. But the number is growing as Muscovites make major changes in their lifestyles, just as their country is making major changes in its economic and political system.

GEOGRAPHY TASKS

1 Imagine you live in Moscow. Write a letter to a friend describing what makes life so attractive in Moscow as well as the problems you experience.

2 Collect information about Moscow from a travel agent, a tourist information office, or the Internet. Plan a week's visit to Moscow including how you will get there, where you will stay, and about how much it will cost.

3 Compare Moscow with a large West European city you are familiar with. How is Moscow particularly different?

......Japan

FROM SILK TO SILICON CHIPS

Japan doesn't seem like a very likely country to become a modern industrial giant. It has very few natural resources, hardly any coal or oil. Earthquakes, volcanic eruptions, typhoons, and huge destructive waves called tsunamis are all relatively frequent. Although Japan is about as large as Finland, most of its 125 million people as well as most of its industrial and farming areas are crowded along the coastal regions because 80 per cent of the country is too mountainous for farming or industry.

Yet in spite of these problems and the fact that its factories were in ruins after the Second World War, Japan's economy has grown so fast that the country that used to be known for making silk cloth and samurai swords is now a leading producer of high-tech electronic products as well as steel, ships and cars. Japan is now the second most powerful industrial country in the world.

In the 1950s and 1960s the emphasis was on building up heavy industry such as steel and ship-building. A second phase began in the 1970s with a boom in consumer electronics and motor vehicles. In order to compete in the international market, and because of its limited space and resources, Japan concentrated on making things that were smaller, cheaper and more efficient. High value-added products such as transistor radios didn't need a lot of raw materials to make. Their value came from the skill and time needed to put them together. Japan also led the way in producing cars that used fuel more efficiently. In the 1980s the focus was on knowledge-intensive products, such as computers and biotechnology, that need huge amounts of money for research and development. Recent growth areas in the Japanese economy are banking, tourism and recreational activities.

Figure 1 **Japanese products on the world market (1994–1995)**

2	1	3	
USA	Japan	Germany	Cars
South Korea	Japan	Germany	Ships
USA	Japan	China	Steel
Japan	USA	China	Paper
Germany	USA	Japan	Plastics
Taiwan	USA	Japan	Synthetic fibres
China	USA	Japan	Refrigerators

GEOGRAPHY TASKS

1 Make a list of Japanese companies you know and the products that they make. What generalisations can you make about the types of products that are "made in Japan"?

2 Use the information from Figure 2 to draw a divided bar graph (see the example). Arrange the information in ascending or descending order.

Figure 2 Japanese goods: percentage used for the home market and for export

Goods	Washing machines	Microwaves	Pianos	Bicycles	Watches	Cameras	Calculators	Colour TVs	Refrigerators	Motorcycles	Batteries	Video recorders
Percentage exported	32	78	29	13	83	82	76	60	25	46	29	82
Percentage used for home market	68	22	71	87	17	18	24	40	75	54	71	18

Example

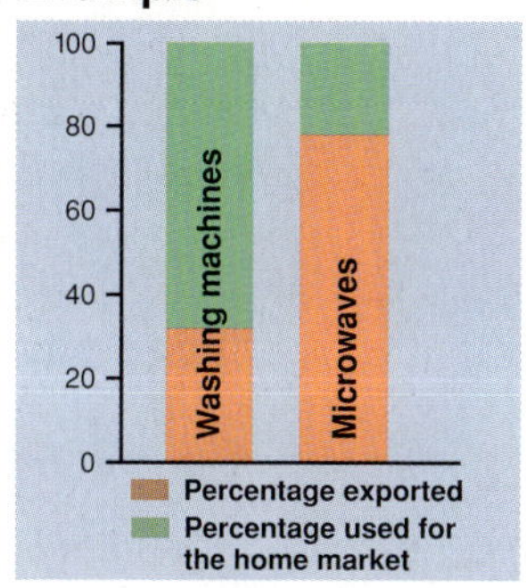

3 Figure 3 shows employment changes in Japan and Germany.

a How did the three economic sectors change between 1980 and 2000 in Japan?

b Can similar changes be seen in Germany's employment structure?

c What does it mean to the individual person that the number of people working in the primary sector has dropped since 1980?

Figure 3 Changes in employment by sector (in %)

	Japan			Germany	
Economic sector	1980	1996	2000	1970	1999
Primary (agriculture)	11	5	5	9	3
Secondary (manufacturing)	35	34	30	49	33
Tertiary (services)	54	61	65	42	64
Total	100	100	100	100	100

PRACTISE YOUR ENGLISH

The following words and expressions will help you to describe changes in employment by sector

Employment in the primary/secondary/tertiary sector …
… rose / grew / declined / went down by … per cent between … and …
There has been a rapid/slow increase/decrease in the percentage of people in the … sector.

PRACTISE YOUR ENGLISH

These words and expressions will help you formulate statements about Japanese products.

In my experience, Japanese products are generally / are often / tend to be / are known for being
cheap/(in)expensive/(in)efficient
well/poorly made
of low/reasonable/high quality

Whereas in 19… cotton fabrics were important / the most important export good, in 19… they had risen/increased/fallen/decreased/declined in importance.
In 19… the emphasis in export goods shifted from … to … .
Motor vehicles rose/increased/fell/decreased/declined in importance between 19… and 19… / were not as important as Japanese export goods in 19… as they were in 19… .

Figure 4 **Japan's five most important export goods**

	1960	1975	1984	2001
1	Iron and steel	Iron and steel	Motor vehicles	Motor vehicles
2	Cotton fabrics	Motor vehicles	Iron and steel	Semiconductors
3	Ships	Ships	Cassette recorders	Office machinery
4	Clothing	Metal goods	Office machines	Chemicals
5	Fish products	Precision optics	Ships	Metal goods

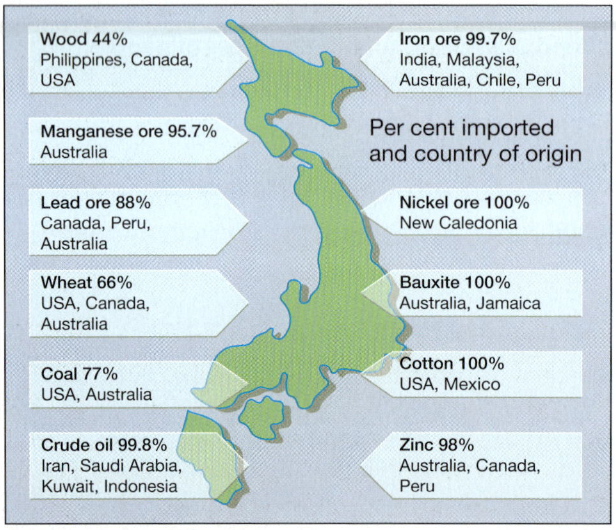

Figure 5 **Japan's dependence on raw materials (1994)**

Figure 6 **Ranking according to gross national product (GNP)**

	1958	GNP in billion US $	1995	GNP in billion US $
1	USA	454	USA	7100
2	Great Britain	65	Japan	4964
3	Germany	58	Germany	2252
4	France	50	France	1451
5	Canada	34	Great Britain	1095
6	Japan	32	Italy	1088
7	Italy	29	China	745

GEOGRAPHY TASKS ■■■

1 Use the information in Figure 4 to answer these questions:

a) Which two of Japan's exports in 1960 were heavy industries?

b) What had happened to the importance of these two industries by 2001?

c) How would you describe Japan's top five export industries in 2001?

d) Explain why Japan has changed its industrial production in these ways.

2 Japan is rich and poor at the same time. Use information from Figures 5 and 6 to explain this fact.

3 Using an atlas and the information in Figure 5, mark on a copy of a world map the sources of Japan's raw materials. Draw a line connecting each source area with Japan and label each line with the raw material being moved.

4 Which countries is Japan most dependent on for its raw materials?

5 Look at Figure 6. For each country, divide the GNP for 1995 by the value for 1958 to find the factor of increase for that period. Which country shows the largest rate of increase? How does this compare with the rate of increase in other countries?

WHY HAS JAPAN'S INDUSTRY BEEN SO SUCCESSFUL?

The Japanese have a saying: "The nail that sticks up is hammered down". In Japan, people are taught never to be "nails that stick up" but to blend in and be absolutely loyal to their group. This group can be the family, school, firm, or all of Japan. One theory suggests that this way of thinking dates back to early Japanese history when they first began to grow rice. Everyone in a village had to work together to build the complicated system of waterways needed for rice to grow. Whatever its origin may be, the Japanese value of working hard for the good of the firm and for the good of all of Japan is just one reason for the country's economic success.

Growing up Japanese

"Examination hell" is what the Japanese call the time when they are preparing to get into high school or university. Only those with the best entrance examinations can get into the best schools and only those from the best schools have a chance of getting into the best universities and later getting the best jobs. In addition to their regular schools, about 80 per cent of all Japanese pupils go to "cram-schools" called *juku* where they spend many hours a week, sometimes both early in the morning before school and late in the evening, preparing for the entrance examinations. There are even special schools to prepare students to get into the best *juku*. Everyone in Japan must go to six years of elementary school and three years of junior high school. About 90 per cent also go on to high school for three years. Most then complete a two- or four-year university course before taking another entrance examination to get a job with a firm.

Mid-term examinations in a Japanese high school

Japanese students are expected to do everything as a team and for the good of the whole group. The skills of expressing an individual opinion and debating issues are not valued highly in Japanese education. Learning how to work in harmony in a group is more important. In school as at the workplace, everyone older than you is *Sempai*. That means they are expected to help you and you are expected to respect them. Everyone younger than you is *Kohai* and they are to respect you and you are to help teach them.

Working in Japan

Big Japanese firms are often compared to big traditional families. The firm is expected to take care of its employees and give them many benefits and a job for life while the workers in turn are expected to give their complete loyalty to the firm and work as hard as they can for the good of the company. In some companies, this strong group feeling is expressed through a company song that all the employees sing together every day before starting work.

While the very large companies with their high-tech production lines and family-type ties are the most well-known, 70 per cent of Japanese employees actually work in small or medium-sized firms. Many of these firms make parts needed by the large firms. Jobs here are much less secure, especially since it is becoming cheaper to have parts made in other Asian countries where labour costs are lower.

Japan is still a man's world; there are very few women in any leadership positions. Differences in wages between men and women are greater in Japan than in any other industrialised country. Paying women lower wages is one way Japan can keep its labour costs down. Even women with university degrees earn much less than men, who still mainly expect them to smile and serve them tea. A large number of Japanese women invest most of their time not in a career of their own but in the school careers of their children, deciding which *juku* to send them to, and sometimes even going to classes themselves if the children are ill.

Junichi Watanabe *is a salaryman, the term used in Japanese for a businessman in a large firm. Although he could take the train to work, he prefers his own car, even if it takes him two hours to get to work in the morning because of the heavy traffic. At the end of the day, he wouldn't think of leaving the office until the boss first stands up and is ready to leave. He often doesn't get home to a hot meal with his wife until about 10 pm because he goes out with his colleagues for snacks and sake and a visit to a few karaoke bars. Japanese employees officially work only 39 hours a week, but Mr Watanabe, like most Japanese people, is at his job at least 43 hours a week, plus all the time he is expected to spend after work with his colleagues. He officially has 40 days of paid holidays, including public holidays, but in the last six months, he only took two days off, and those were on public holidays. Like many Japanese employees, he doesn't take all the holidays he could. He has been working for the same company since he finished university and knows he will be able to work there until he retires. He will probably spend his last years as a "window-looker", the Japanese expression for older employees who can't be fired but who don't have much to do besides looking out the window.*

Technology at work

Since the end of the 19th century when Japan wanted to grow as quickly as possible from a poor agricultural country to an industrial power, Japanese people have travelled to Europe and the United States to learn about production techniques and technology and then have adapted and often improved those ideas. Large amounts of money are invested in research to find ways to produce things faster and more efficiently. Industrial robots, for example, are used in Japan more than in any other country.

	Japan	USA	Western Europe
Labour hours per car	16.8	25.1	36.2
Assembly mistakes per 100 cars	60	82	92
Suggestions for improvement per employee	62	0.4	0.4
Absenteeism in per cent of employees	5.0	11.4	12.1
Employees working in teams, in per cent	69.3	17.3	0.6
Hours needed to train a new employee	380	46	173

Figure 7
Comparing car competitors

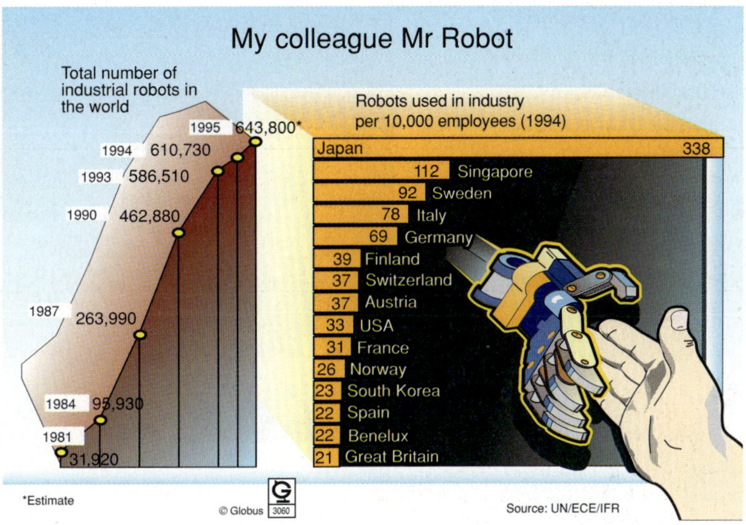

Figure 8
My colleague Mr Robot

PRACTISE YOUR ENGLISH

You can use the following expressions when describing working conditions and the use of technology in industry.

Japan/Spain/Austria uses twice as many / ten times as many robots as ... .
Japanese car firms invest about x times more / x more hours in job training than companies in ... .
It takes more/fewer labour hours to make a car in ... because more/fewer robots / more/less automation are/is used.
In Japan it takes ... hours to build a car, but in Western Europe it takes ... hours.
The percentage of employees working in teams is higher/lower in ... than in ... .
In ... the workers appear to be more/less efficient/lazy/enthusiastic/involved / better/worse trained than in ... .

GEOGRAPHY TASKS ▪▪▪

1 Interview one of your parents and write a short report about his or her working conditions. You could ask about working hours, holidays and job security, for example.

2 Match the heads and tails below to make five sentences:

Heads	Tails
1 Robots can do repetitive jobs at greater speed, which …	a … the use of high technology robots.
2 Employees work in teams, and loyalty to their team …	b … make more suggestions for improvement.
3 Because workers want to do their best for their work team they …	c … reduces the number of labour hours needed to make a car.
4 By training workers very well …	d … the number of assembly mistakes is reduced.
5 The amount of training required by workers is increased by …	e … reduces absenteeism.

3 In what ways are Japanese car factories more efficient than those in the USA or Western Europe?

4 How is the role of women in Japan different from their role in Germany?

MAKING ROOM FOR GROWTH

As most of the country is so mountainous, finding enough flat land for large industrial areas is a problem in Japan. In the 1960s the government began working on plans to make new land by filling in parts of the sea. About 1500 km^2 of new land has been reclaimed this way.

One of these projects, called Mizushima ("Water Island"), was built in the bay of Kurashiki. Because rivers brought so much sand and rubble into the bay, the water there was only about 5 m deep, even as far as 6 km from the coast. Dikes were built to block the water and then stones and soil from nearby hills were dumped into the bay. Shipping lanes and an 18 m deep harbour-basin were dug so that large ships and tankers could use the new harbour next to Mizushima. The sand and gravel that were dug out were used to make new land, as well as waste material from factories and construction sites that was also dumped into the sea. By 1985 about 2500 ha of land had been reclaimed. Most of it is about 2.7 m above sea level. Nearly 50,000 people work on the new "Water Island" in more than 90 large firms such as steelworks, petrochemical works, oil refineries, car assembly plants and engineering works.

Although there had been plans to build many more new industrial areas such as Mizushima, progress has slowed down as heavy industry is not growing as fast as once expected in Japan. Newer high-tech industries such as those that produce medical-electronic machines, laser machines and computers do not need so much room. And because the finished products are not so large and heavy, it is not so important that they are built near the coast to save on transportation costs.

Reclaiming land has also created new problems. The coastline has been shortened, meaning fewer swimming beaches and recreation areas as well as the destruction of biotopes. Air and water pollution in the new industrial areas is also a problem.

Figure 9 **How land is reclaimed**

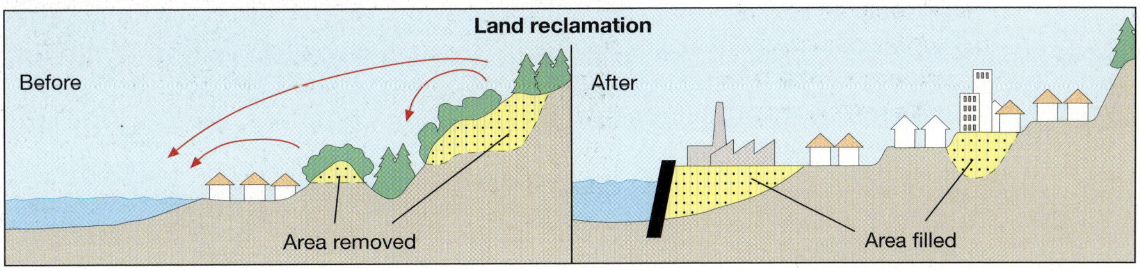

PRACTISE YOUR ENGLISH

You can use these expressions to describe the map showing Japan's industrial areas.

Most of the industrial areas are concentrated around the cities of ...
/ are located along the west/east coast / on inland seas.
... per cent of all Japanese industry is concentrated in the area near ...
The following words and expressions will help you describe how land is reclaimed:
... is/are built/constructed ...
... was/were taken out / removed ...
... has/have been levelled ...
put in ...
dug (out) ...
replaced ...
cleared (away) ...
dumped ...

GEOGRAPHY TASKS ■■■

1 **Explain why the site near Kurashiki was chosen as a place to reclaim land and build a new industrial area.**

2 **Using information from the text and Figure 9, describe how new land areas are reclaimed.**

3 **Using an atlas, draw an outline sketch of Japan and fill in the names of the four main islands, cities with more than one million inhabitants, and the town of Kurashiki. Indicate the main as well as smaller industrial areas.**

4 **Describe the locations of the industrial areas and find three reasons for this particular pattern. Discuss advantages and disadvantages of this pattern of location.**

IMPORTS, EXPORTS, INVESTMENTS

One key to Japan's economic success is the high productivity of its industries. Another key is international trade. As an isolated island country with few raw materials, Japan is completely dependent on trade with other countries, especially the United States, its most important trading partner. In 1996 Japan exported goods worth 44.7 billion yen while importing goods worth 38 billion yen. The difference between the value of imports and exports is called the trade balance. Japan has a trade surplus; it sells more goods abroad than it buys.

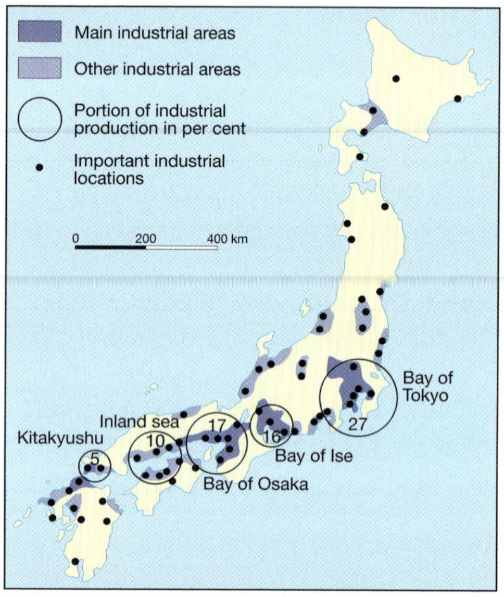

Figure 10 **Main industrial areas**

Japan and Europe

The European Union is a large and rich market for Japanese products, but the cost of transporting Japanese products to Europe is high and besides that, European countries have high import tariffs on goods made in other countries. One way to solve these problems is for Japanese firms to build factories in Europe and make the products near the markets that will buy them.

The United Kingdom is especially attractive for Japanese companies. Wages there are lower than in many other European countries and there are many skilled but unemployed workers eager to work in the new factories. Because these new factories mean jobs and investment money, communities are usually glad to welcome the investors and the British government gives them special tax incentives. Nissan, for example, has built a car factory in Sunderland that produces 200,000 cars a year and gives jobs to 3500 workers.

Of all the European Union countries, Germany has the most trade with Japan, about a third of all of Japan's trade with Europe. Germany imports motor vehicles, computers and memory circuit boards as well as precision optics from Japan. It exports mostly motor vehicles, medical products and equipment, vehicle parts and printing machines. Germany has a trade deficit with Japan; that is, it imports more than it exports. In the 1990s, this deficit began to be reduced with German exports increasing and Japanese imports decreasing.

About a thousand Japanese businesses have branches in Germany including many manufacturing plants. Looking in the other direction, about 440 German firms have branches in Japan. Most of them are in the distribution sector.

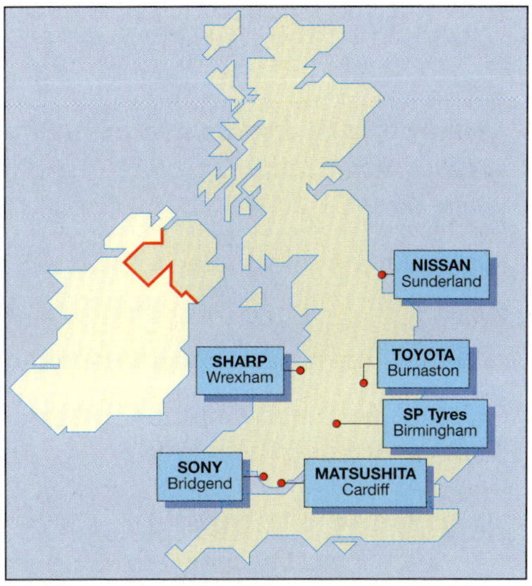

Figure 11 **Japanese firms in the UK**

Little Tokyo on the Rhine

With about 8000 Japanese people and 450 Japanese firms in the area, Düsseldorf has become the centre of Japanese life in Europe. There are some places in Düsseldorf where you can go to a restaurant, bank, hotel, hairdresser or doctor without having to speak a word of German.

Although there had been some ties between businesses in Nagasaki and Düsseldorf in the 19th century, the modern Japan-Düsseldorf connection began in the 1950s when Japanese businessmen started coming to Germany. Japan wanted to build up its steel industry and needed machines and other German industrial products as well as know-how and technology. Düsseldorf, known as the "office desk of the Ruhr area" was an ideal location for this. It is also near the ports of Rotterdam and Antwerp and is the centre of a very populous area in Germany, which offers a big market for Japanese products.

As Japan's industry focused more and more on electric and electronic products and needed new markets for them, those businesses as well as office machine companies, banks, insurance and distribution firms came to Germany. About 30,000 Germans work for Japanese companies in the Düsseldorf area.

Many of the Japanese business people stay in Germany for about two to five years. They bring their families with them and want to keep up their Japanese culture and traditions. As a result, a whole Japanese community has developed with its own Japanese schools, a kindergarten, temple (photo above), cultural centre, as well as many shops and restaurants which make "Little Tokyo" the centre of German-Japanese business and cultural ties as well as a tourist attraction.

PRACTISE YOUR ENGLISH

Draw a mind map using the words and expressions relating to trade that are found on pages 42 and 43. Add other words and expressions from this word field that you already know.

GEOGRAPHY TASKS

1 You work for a Japanese firm that makes motorcycles. Explain why you would like to build a manufacturing plant in Europe and where you would like to build it.

2 Describe the trade balance between Germany and Japan. Who is benefitting most from this trade balance at the moment?

3 a) Find three reasons why Düsseldorf became known as Little Tokyo.

b) What are the advantages to Germany of this Japanese investment?

GROWTH AT WHAT COST?

"It was around 1954 when I moved to Nishiyodogawa. In those days the area was a greyish, dirty place, and there were many dead fish floating belly-up in the river. When I hung white sheets out to dry, they would soon turn black. After living there a while I too became a victim of pollution. People who have never experienced the suffering of heavy coughing caused by pollution will find it hard to imagine. No matter how much you cough, it just doesn't stop. Your throat is stopped up by thick phlegm, and finally you're not able to breathe any more. [...] Sometimes I think that I'd be better off dead because it would set me free of this. Still, I keep going because I don't want our children and grandchildren to suffer like this." Tomoe Yamaki, member of the Japan Air Pollution Victims Association, 1994

After World War II, Japan's main interest was economic growth and building up its heavy industry as fast as possible. By the 1960s Japan was becoming a strong industrial country, but at the same time it was one of the most polluted countries in the world. The cities were filled with a milky white smog and many people wore face masks when they went outside. People were shocked when strange new diseases developed and hundreds of people died because of pollution in the air and water. Groups of pollution victims and their friends began to organise and put pressure on the government.

Finally in 1967 the government was shamed into action and they passed the first Pollution Countermeasures Basic Law. They established the first environmental quality standards, began pollution-control programs, and started to help victims of diseases caused by pollution. In the 1970s the Polluter Pays Principle forced factories to accept financial responsibility for damages they caused to the community. Another big step was a law limiting the production of chlorofluorocarbons, of which Japan was a major producer, following international warnings about the destruction of the ozone layer.

Because much of the pollution in Japanese towns and cities is from cars, some cities do not allow certain kinds of cars to enter the central districts, and in some places no cars are allowed. Japan spends 2 per cent of its GNP on pollution control, more than any other country in the world. Being a country surrounded by water, Japan has a strong interest in doing its part to fight the global warming of the atmosphere. While the environmental situation is not as dramatic as in the past, the problems are far from solved. Many foreign visitors notice, for instance, that many Japanese people still have a strong throw-away consumer mentality and recycling is a relatively new idea for them.

Although its pollution laws are strict, Japan as well as much of East Asia still has a big problem with acid rain destroying its trees and polluting its water. Japan's industry says it is mostly caused by China's coal-burning factories and power stations. Although there are many pollution laws in East Asia, many of them are not enforced, and there is a lack of coordination among the different East Asian countries.

Figure 12 Some sources of air pollution in Japanese towns and cities

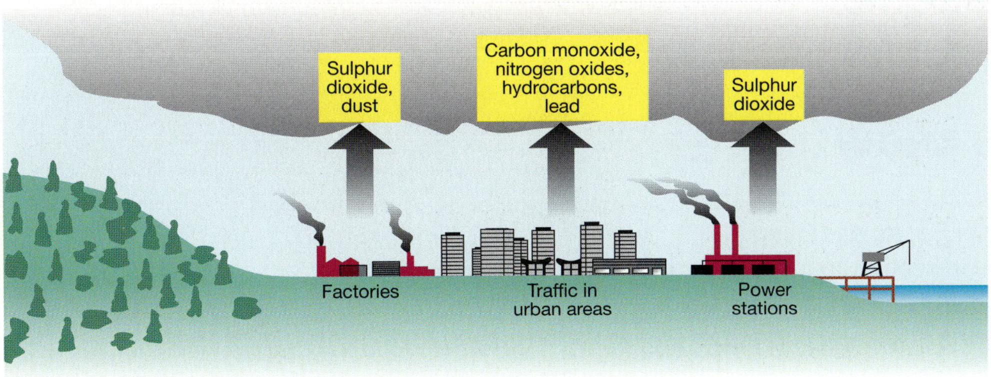

GEOGRAPHY TASKS

1 What are the main sources of air pollution in Japan? What about in your own region? Make comparisons.

2 Find four specific environmental problems caused by air pollution.

3 Identify three laws which the Japanese government has passed to reduce air pollution.

Figure 13 The system of environmental protection in Japan

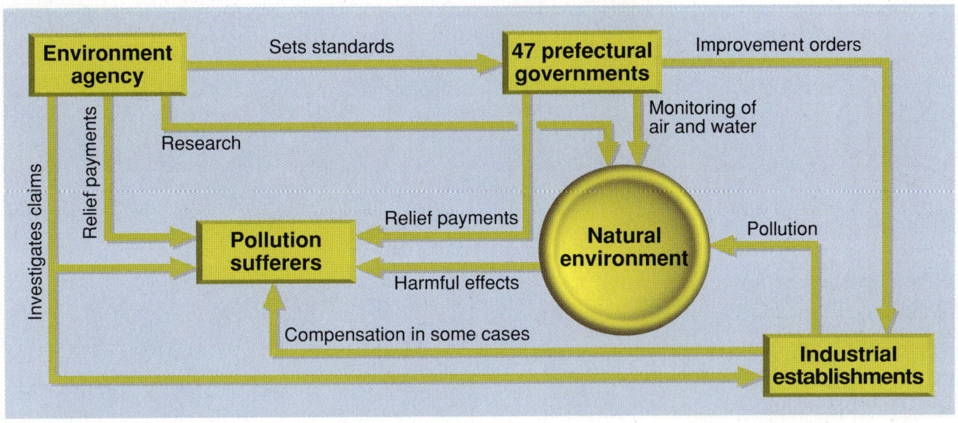

4 Answer the following questions using information from Figure 13.

a The most important part of the diagram is the role played by the government's Environment Agency. What four things does this agency do?

b The 47 prefectures are rather like German *Landkreise.* What three things do they do?

c If you thought that you were suffering from the effects of industrial pollution in Japan, who would you complain to?

d If you were the factory causing the problem, what action might be taken against you?

e Working in a small group, decide how each of the following groups of people might also help to control industrial pollution in Japan: shareholders, trade unions, customers, schools.

5 What causes acid rain and why has it been difficult to solve the problem in East Asia?

...... China

DID YOU KNOW ...?

- China is the third largest country in the world after Russia and Canada.
- Over 1.2 billion people, about one-fourth of all the people in the world, live in China.
- Although China is very large, most of the people live on only 15% of the total land area.
- The Chinese have had a common culture for longer than any other group of people in the world. They were the first to make paper, ink, silk, porcelain, fireworks and gunpowder.
- The capital of China is Beijing. In an older system for writing Chinese words in Roman letters, the city is called "Peking".
- In 1990 there were only 240,700 privately owned cars and buses in China. In 1996 there were over 1.4 million.
- In only one year, from 1995 to 1996, the number of telephones in China jumped from 6.3 million to 70.5 million.

Figure 1 **Infant mortality**

Number of deaths per 1000 live births, 2001	
China	**16**
Germany	5
India	63
Japan	4
Mexico	25
Russia	20
S. Africa	60
USA	7

Figure 2 **Population shift**

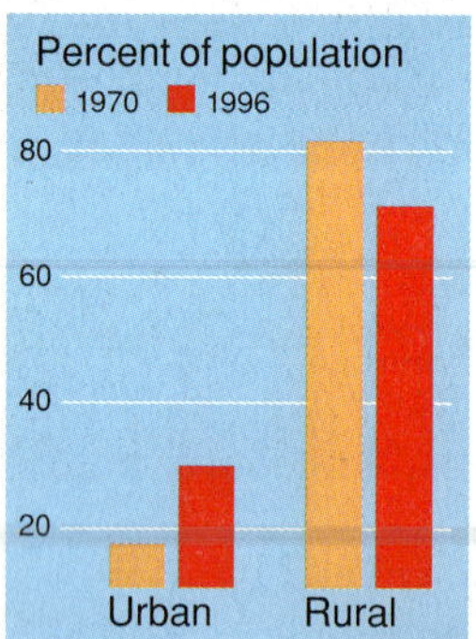

Figure 3 **Production**

Gross National Product per capita, 1996, US $	
China	**750**
Germany	28,870
India	380
Japan	40,940
Mexico	3670
Russia	2410
S. Africa	3520
USA	28,020

Figure 4 **Water pollution**

Emissions of organic pollutants, thousands kgs/day	
China	**7396**
Germany	811
India	1664
Japan	1469
Mexico	143
Russia	1615
S. Africa	242
USA	2585

Chinese mini-glossary

bei (pe)	north
dong	east
hai	sea
he	river
hu	lake
huang	yellow
jing (king)	state capital
nan	south
shan	mountain
shang	exit
xi	west

What do these Chinese place names mean?

Nanjing
Beijing
Shanghai
Huanghe
Shanxi
Shandong
Hubei
Hunan

Yellow China Yellow is for the colour of the wheat that grows in this region as well as for the Hwang Ho, the "Yellow River", that carries large amounts of yellow soil called *loess* from the hills of the west and leaves these sediments in the lowlands. The route of the river is always changing and often floods the land. The people try to protect themselves and their land by building dams and channels. The land is fertile and there is also coal and iron ore for industry.

Green China About half of all Chinese people live in the fertile basins and plains of the southeast. The colour green is for the green rice fields. Rice, the most important food in China, is grown on terraces on the slopes of the hills. Also typical for this region are narrow canyons and tall limestone towers in unusual shapes. This landscape has been a popular theme in traditional Chinese art.

Dry China Northwestern China has high mountain ranges that surround great sandy basins where very little rain falls. Although there are rich mineral resources here, there is no industry. The only people who live in the desert areas are some non-Chinese nomads. There are a few larger settlements and irrigated farmland in river oases.

Cold China Tibet is an isolated area of great mountain ranges and endless high plateaus, an area called The Roof of the World. In 1950 China conquered Tibet and many Chinese people settled in the larger towns. Many Tibetans are semi-nomads who raise yaks. They make clothes and blankets from the yaks' fuzzy fur to protect themselves from the cold. They also get milk and meat from the yaks.

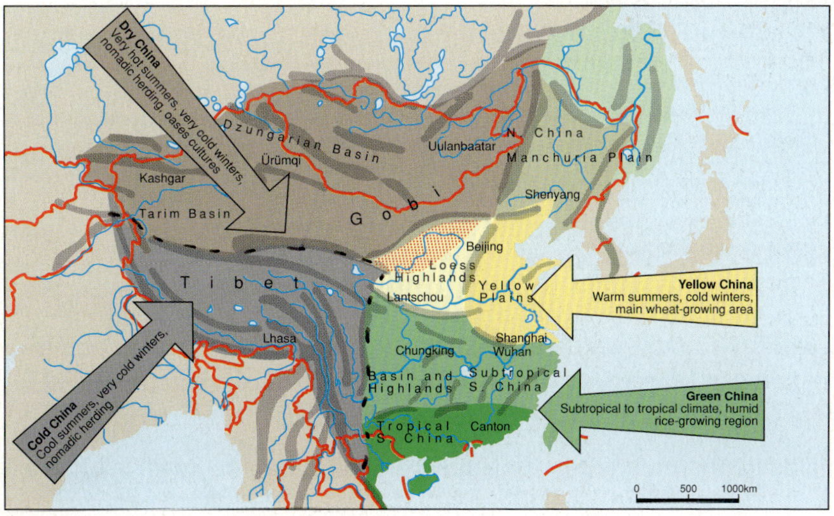

Figure 5 **Land use in the four regions**

PRACTISE YOUR ENGLISH

You can use the following expressions when discussing soil fertility, precipitation and distribution of population.

The most / least fertile soil is in the north / south / near the population centres / cities.
The driest / wettest part of China is in the west / east / southeast.
Population is very dense / quite sparse in the west / east / along the coast / in the desert areas.
Areas where the population is most / least dense are …

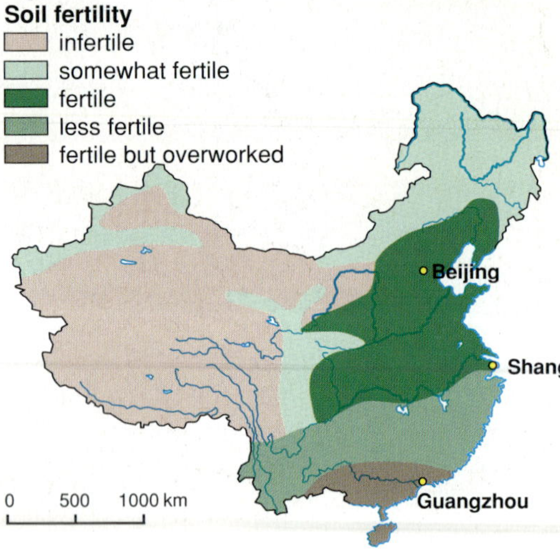

Figure 6
Soil fertility

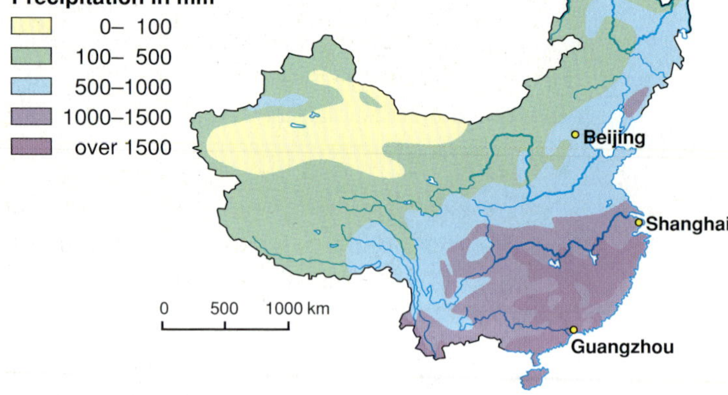

Figure 7
Precipitation in mm

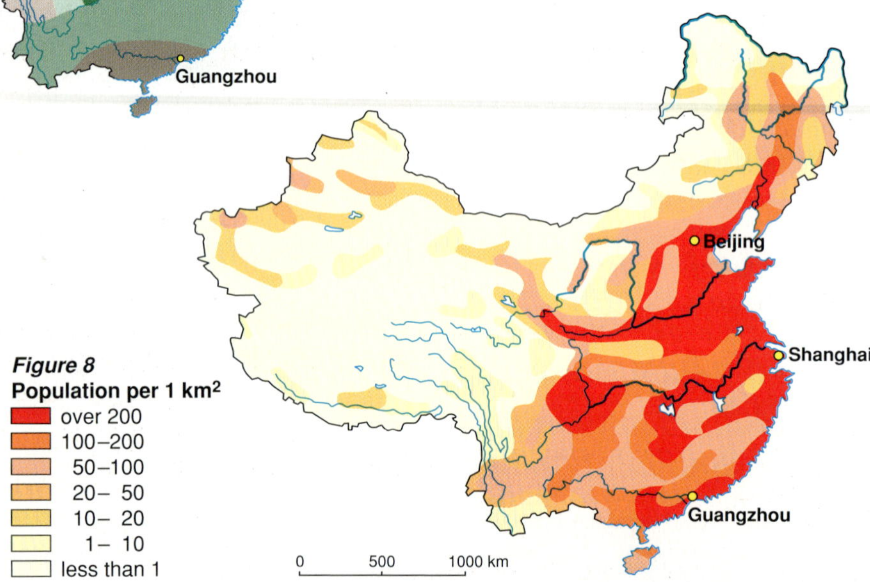

Figure 8
Population per 1 km²

GEOGRAPHY TASKS

1 **Using information from the text and the maps, make a table with facts about the different faces of China. Use these headings: Type of Landscape, Relief, Climate, Land use, Soil fertility and Population density.**

2 **What problems might the unequal distribution of population shown in Figure 8 lead to?**

3 **What similarities are there between the three maps in Figures 6, 7 and 8? Do these similarities suggest a link between soil fertility, precipitation and population density? Explain your answer.**

4 **What other factors might explain the pattern of population density?**

THE YANGTZE RIVER AND THE THREE GORGES DAM

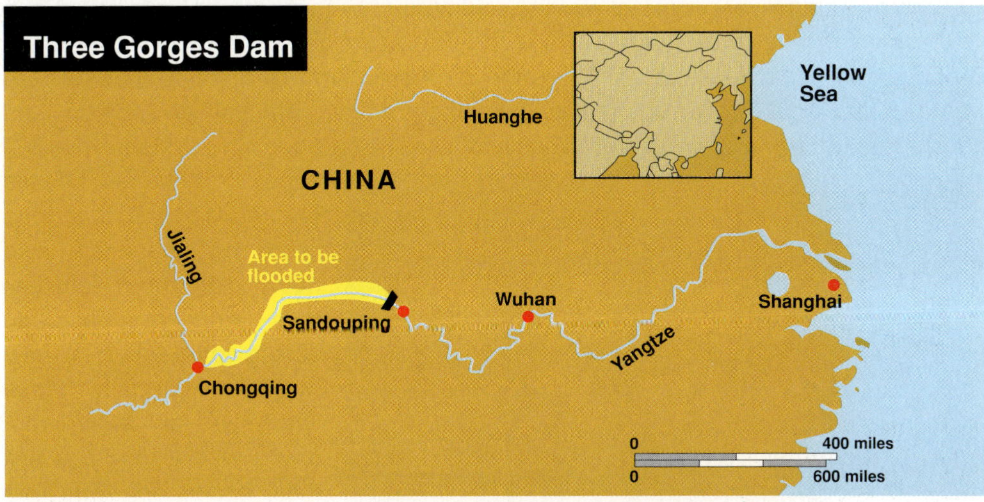

People are already calling it the eighth wonder of the world. It is the largest engineering project the world has ever seen and one of the most controversial. If the Three Gorges Dam is finished as planned in 2010 it will be the largest, most expensive and most powerful dam and hydroelectric power station in the world. The dam will be 185 metres high and stretch more than a kilometre between the two banks of the Yangtze River. It will create a deep lake 600 kilometres long.

Reasons for the dam:
- To control the Yangtze River, the longest river in China (nearly 6000 km) and source of many floods which have killed hundreds of thousands of people
- To produce hydroelectric power for China's growing industry. It will produce about 100 billion kilowatt-hours of hydroelectric power each year. This will reduce air pollution, acid rain and global warming because China will not need to burn so much coal for energy, now the source of three-fourths of its power. Chinese engineers believe the dam will create the same amount of electricity as ten nuclear power stations.
- To make the Yangtze River into a smooth waterway for large ocean-going ships which will be able to travel 1500 km up the river.

But environmental groups warn of many problems:
- At least 1.2 million people will have to leave their homes because their land will be flooded. Thirteen cities, hundreds of small villages, and thousands of factories and mines as well as many important old cultural sites will be under water.
- Fish and other wildlife will be threatened by the changed environment.
- Farmland will no longer receive the water and silt it needs.
- Water will not be able to flow naturally and will become polluted. Tons of sediment will become trapped behind the dam's turbines and may damage them.
- The dam may be on the top of an earthquake fault line. There is also a danger of landslides.
- The dam will destroy some of China's finest scenery, which is also an important source of tourism income.

CHINESE AGRICULTURE – A SUCCESS STORY?

Throughout Chinese history there have been different systems of land ownership. The communist government which took power in 1949 organised all the small farmers into communes with about 15,000 people each. The land and equipment was all owned by the state and it was not possible for individual farmers to own land or to grow what they wanted. Since the early 1980s this system has begun to change. Under what is called the "responsibility system", the land is still owned by the commune, but individual farmers can now make a 15-year contract with the commune. The farmers can lease some land for themselves. They must agree to sell a certain amount of their crops to the state at a set price and to pay taxes. But if they produce more than that amount, they can sell whatever they want on the free market where prices are set by supply and demand and not by the state. Many farmers have become relatively rich through this system. They can afford to build new houses and have more than one child. But there are also many poor farm families, those who have not been able to adapt well to the new system, especially those who live far away from towns where they can sell their goods and who live where the land is poor.

On the whole, Chinese agriculture has improved greatly in the last years. Since farmers have been able to work independently, they have produced much more than before. China can grow enough food to feed its own people as well as some agricultural products for export, such as tea, spices and vegetables.

Figure 9 **Agricultural production in millions of tonnes**

	1965	1970	1975	1980	1985	1990	1995
Rice	90.7	113.1	128.7	142.9	171.3	191.6	206.4
Wheat	25.2	29.2	45.3	55.2	85.8	98.2	116.1
Maize	23.2	33.1	47.4	62.7	64.1	97.1	106.5
Fruit	3.6	4.7	6.3	8.4	13.2	19.9	25.3
Cotton	2.1	2.3	2.4	2.7	4.1	4.5	5.6

TWO CHINESE FARMERS

The Jin family's village in the Hebei province is only about 200 km away from the magnificent facades of the modern hotels in Beijing. But their primitive peasant huts seem to belong to a different world. Mr Jin has laid out some small terraced fields along the steep mountainside of the Taihang mountains and he has planted a few dozen maize and sweet potato plants. His smallest field is not even as large as the family bed. Altogether he farms on an area of only 1/5 hectare. It was not possible for him to get any more land in this poor mountain village. "What we harvest in one year isn't enough to keep us from going hungry," says the farmer. Jin's daughter had to leave school. She works long hours in a brick factory for only 80 yuan a month in order to buy an occasional sack of maize for the family.
(Adapted from *Geo-Special China*, February 1994)

The Wang family from the Canton area is one of China's farm families that has become relatively rich. Mr Wang has leased a pond and runs a fish-breeding business. His customers are large restaurants in nearby Canton. Mrs Wang took out a loan and built a greenhouse where she grows vegetables and sells them on the free market.
Together, the Wangs earn about five times as much as they did in the commune. The Wang family has built themselves a two-storey house. They proudly show off their refrigerator, washing machine and colour television. Now they are saving their money to buy a used truck so they can transport their products to Canton themselves.
(Adapted from the *Hamburger Abendblatt*, March 1992)

GEOGRAPHY TASKS

1 **Copy and complete this table to show the changes that have taken place in Chinese agriculture:**

	Commune system	Responsibility system
Date introduced		
Who owns the land?		
Who decides what to grow?		
What happens to the crops?		

2 **Look at Figure 9. Draw separate line graphs to show the changes in production of fruit, wheat and cotton. Mark on the graphs a vertical line at 1980, the date when the responsibility system was introduced. Does this system seem to have had any impact on food production? Why might farmers work harder to produce more crops under the responsibility system?**

3 **Suggest three reasons why the Jins have not been able to benefit from the responsibility system.**

CHINESE INDUSTRY – PROGRESS AND PROBLEMS

If you are wearing sports shoes now, they were probably made in China. Most of the toys you played with when you were young were probably made in China, too. "Made in China" has become familiar on products sold in Europe and the North America. About half of all the toys sold in the USA are made in China. The leading American toy company makes most of its products in Asia and makes more of its famous fashion dolls in China than in any other country. Several well-known European and US producers of sports shoes, clothes and equipment also make most of their products in Asia, especially in China.

But these are all recent developments in Chinese industry. Just as there have been big changes in Chinese agriculture since the early 1980s, there have also been big changes in Chinese industry since that time. Before then, all large and medium-sized factories were owned by the state. Prices and wages were also set by the state and production was according to five-year plans. There were also many smaller factories that were owned by the communes and that produced everyday goods. Typical of all Chinese industry was that it was very labour-intensive. Many more jobs were done by hand or with older machines than in other countries. Production levels were often low and the products were of poor quality. China was quite isolated and there was not very much contact or trade with Western nations.

All of that started to change with China's economic reforms in the early 1980s. Four Special Economic Zones were created and China began to experiment with elements of a market economy. Fourteen cities along the coast have been opened for foreign investors. The foreign companies bring money, management skills and modern assembly-line equipment. In China, the costs of labour, land and taxes are low. Most of the products that are made for export are things that even with modern technology still need a lot of hand labour, such as shoes, clothes, toys and electric household goods. Because of the low labour costs, these products can be sold on the world market at prices much lower than when they are made in other countries.

Even though wages for the workers are very low by Western standards and working conditions are often very bad, the booming new industries mean that the standard of living for many Chinese people is rising, especially in the coastal cities. But there are still many problems. There is a growing gap between the rich and the poor, the winners and the losers of the new system. Millions of Chinese people in the farming areas far from the cities have not benefited at all. About 100 million poor and unemployed peasants travel through the country trying to find jobs in the coastal areas. Many of them sleep in train stations or under bridges or in half-finished new buildings until the police throw them out. As some people get rich in new market-economy businesses, others try to get rich fast through crime and corruption, two other growing problems in China.

Figure 10 **Opened coastal areas**

There are four types of firms in China today:

1. *Large firms, for example in heavy industry, are still owned and managed by the state according to centrally organised plans.*
2. *Medium-sized and small firms, for example the textile industry, are organised as co-operatives that are run by managers employed by the cooperative. These firms can make their own decisions about production, distribution of profits, and hiring and firing of workers.*
3. *About 15 million small firms are owned privately, for example transportation firms, repair shops and restaurants.*
4. *Joint ventures between Chinese and foreigners are mostly in the Special Economic Zones and newly opened coastal cities.*

GEOGRAPHY TASKS ■■■

1. **Use your atlas to locate the natural resources that are found in China.**
2. **List the most important industrial centres in China. Are they linked to the areas of natural resources you identified on your map? What else may explain the location of the main industrial areas?**
3. **Develop a diagram to show the organization of Special Economic Zones by grouping the following information in the form of a flow chart:**
   - Inputs from foreign companies
   - Inputs from China
   - Names of four Special Economic Zones
   - Products
   - Sold cheaply on the world market

POPULATION PROBLEMS

For most of China's history, it was important to have as many sons as possible to help on the small farms and to take care of the parents when they were old. There were strong feelings against daughters and a woman had a very low social position if she did not have a son. Girls always went to live with their husbands' families, so the Chinese people said, "Raising girls is like watering other people's gardens".

In modern China, health conditions have improved greatly. Fewer children die young and people live much longer. In fact, between 1950 and 1980 the population doubled. It was clear to the government that with over a billion people, something had to be done to slow down population growth. If not, there would not be enough food, schools, houses, and jobs for everyone. In 1979 the Chinese government started a family planning programme. The goal is for each couple to have only one child. Women have to be at least 21 and men at least 22 years old before they can get married. Contraceptives and abortions are free. If a couple has only one child they receive money from the government, free education and better housing. But if they have a second child, they lose their privileges and have to pay the money back to the government. If they have three or more children, they lose 5 to 10 per cent of their income per child.

The government policy has slowed down the rate of population growth, but in 1990 families in rural areas still had an average of 2.8 children compared to 1.3 children among urban families. If the first baby is not a son, many couples will have more children even if it means that they lose money and privileges.

Age	Male	Female
0 – 4	5.4	4.9
5 – 9	4.6	4.2
10 – 14	4.4	4.1
15 – 19	5.5	5.2
20 – 24	5.8	5.4
25 – 29	4.7	4.5
30 – 34	3.9	3.5
35 – 39	3.9	3.7
40 – 45	2.9	2.7
46 – 49	2.3	2.1
50 – 54	2.1	1.9
55 – 59	1.9	1.8
60 – 64	1.6	1.5
65 – 69	1.1	1.2
70 – 74	0.7	0.9
75 – 79	0.4	0.5
80 – 84	0.2	0.3
85+	0.1	0.1
Total	**51.5**	**48.5**

Figure 11 **Population of China by percentage of age group (1990)**

GEOGRAPHY TASKS

1 **What do you think are positive and negative aspects of the Chinese government's efforts to reduce population growth? What conflicts arise if a couple's first baby is a daughter?**

2 **Would you prefer to be an only child or to have many brothers and sisters? What are the advantages and disadvantages of each situation?**

3 **In China, it is not only the large number of people that is a problem, but the age structure as well. Use the demographic data in Figure 11 to make a population pyramid for China. Describe the pyramid and explain the problems that could result from this age structure.**

...... Nigeria

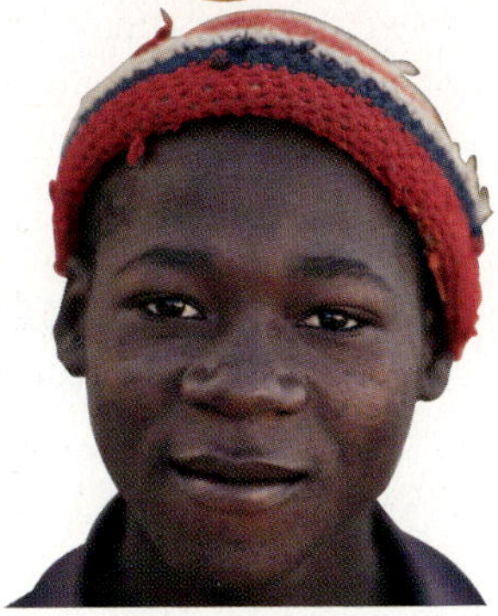

WELCOME TO MY COUNTRY

My name is Ademola and I live in Lagos, the largest city in Nigeria. Over a hundred million people live in my country, more than in any other country in Africa. Nigeria has an area of almost a million square kilometres, about as large as Germany, France and Belgium put together. The name Nigeria is taken from the Niger River, the third longest river in Africa.

My family belongs to the Yoruba tribe, one of the four main ethnic groups in Nigeria, but there are over 200 other groups of people in Nigeria. Each group has its own language, customs and traditions. It is difficult for all these different groups to live together peacefully and so Nigeria's history has been full of conflicts. Europeans first came to Nigeria in the 15th century looking for raw materials and new markets. There was a lot of trade with Europe, including thousands of Nigerians who were sold as slaves and taken to America to work. Nigeria was a British colony from 1861 to 1960 when it became an independent country.

Like many Nigerians, my parents moved to the city from a small village because my father found a job in one of the new factories. Our house is modern and quite nice, but many people in Lagos and other cities live in very crowded slums. Today about half of the people in Nigeria live in towns and cities, but before 1960 most Nigerians lived in small villages and worked on small farms. Along the coast where the climate is tropical, traditional Nigerian houses are made of bamboo sticks with the roofs made of bamboo leaves. In the middle part of the country, houses are made of mud with grass roofs, and in the dry areas of the north, people have mud houses with flat mud roofs.

Our Nigerian flag is green and white. Green is for agriculture and forestry. Agriculture is very important for Nigeria. We export cocoa, rubber and palm oil. We used to export a lot of wood from our tropical rainforests, but too many trees were cut down and now there are not so many left. It is very good for our country that we have many natural resources: coal, iron ore, and of course the most important thing, oil. The white band in our flag is for peace and unity, which is also my hope for the future of Nigeria.

GEOGRAPHY TASKS ▪▪▪

1 **Draw a rough outline sketch of West Africa and mark the position of Nigeria and its neighbouring countries.**
2 **Using information from Ademola's report, an atlas and your own knowledge, draw up a table comparing Nigeria and your own country. Use the following headlines for the rows in your table:**
population, area, history, ethnic structure, urbanisation, natural resources.

Figure 1 **Map of Nigeria showing the geographical regions**

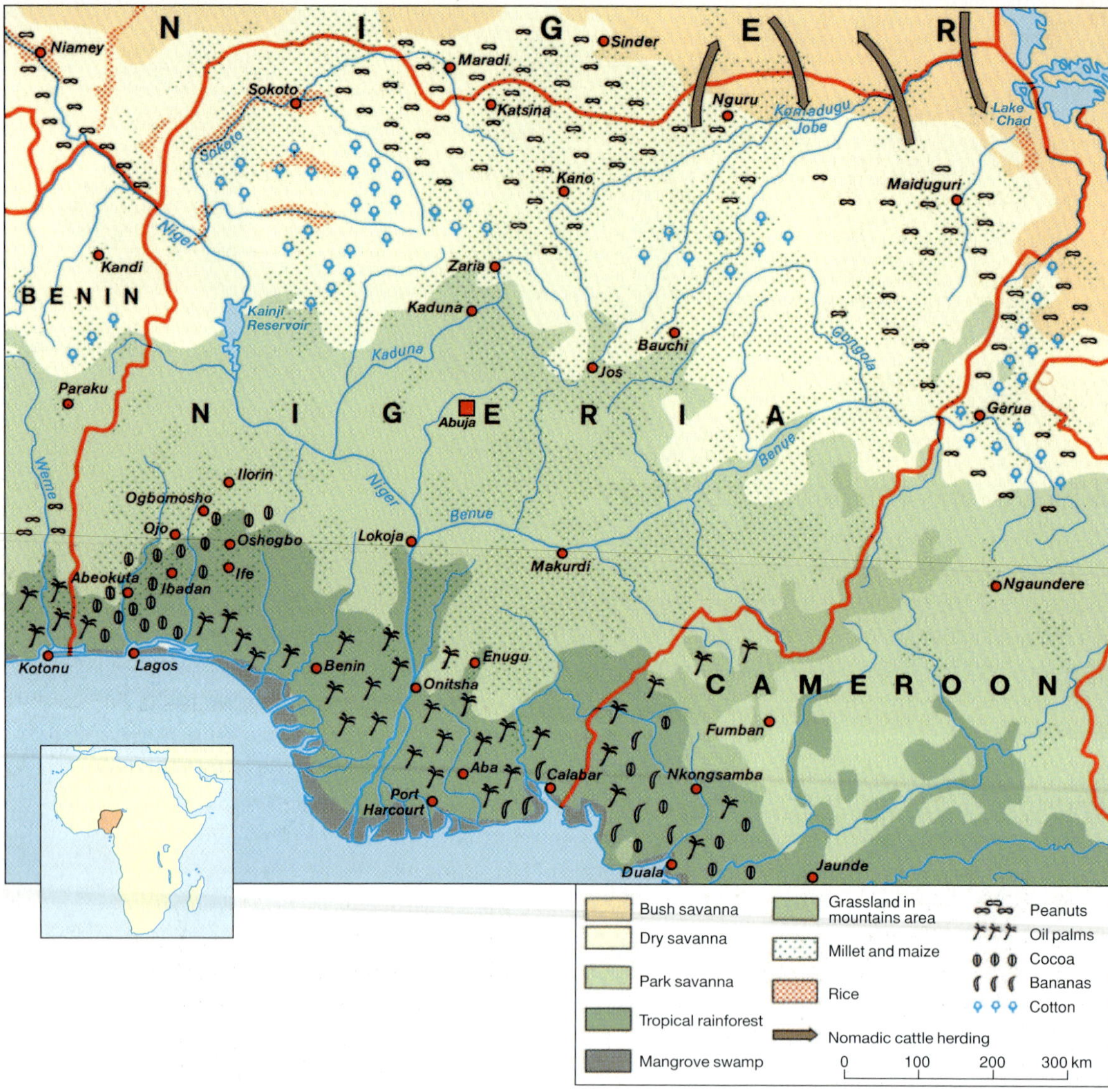

PRACTISE YOUR ENGLISH

These expressions can be used when discussing areas on a map:

Along the southern coast there is a belt / region / zone / area of mangrove swamps.
The middle belt / rainforest area / savanna ranges from … to … kilometres wide.
… and … are grown / raised / the main crops in the … belt.
The farther south/north one goes, the drier / wetter / more humid / hotter the climate becomes.
Compared to the … belt, the … belt is (more) … .

GEOGRAPHY TASK ■■■

Make a chart as shown here with information about the geographical regions of Nigeria. Use the map in Figure 1 as well as the climate data below. Compare the three main regions.

Zone	Natural vegetation	Climate	Main crops
Northern belt			
Middle belt			
Southern belt			

(A) Lagos, Nigeria 6° N / 3° E

	Jan	Feb	Mar	Apr	May	Jun	Jul	Aug	Sep	Oct	Nov	Dec	Year
°C	27	28	28	28	27	26	25	25	26	27	27	27	27
mm	40	57	100	115	215	336	150	59	214	222	77	41	1625

(B) Makurdi, Nigeria 8° N / 9° E

	Jan	Feb	Mar	Apr	May	Jun	Jul	Aug	Sep	Oct	Nov	Dec	Year
°C	27	29	30	30	28	26	26	26	26	26	27	26	27
mm	5	10	20	124	226	193	185	193	279	150	18	2	1407

(C) Maiduguri, Nigeria 11° N / 13° E

	Jan	Feb	Mar	Apr	May	Jun	Jul	Aug	Sep	Oct	Nov	Dec	Year
°C	23	25	29	32	33	31	28	26	27	29	26	23	28
mm	0	0	0	7	38	66	186	221	102	19	0	0	639

AGRICULTURE IN NIGERIA

Traditionally, Nigerians have been subsistence farmers; that means they grow just enough food to feed themselves and their families. Today over half of the people in Nigeria still work in agriculture, and small family farms account for 90 per cent of Nigeria's agricultural production. These farms usually have about one or two hectares of land and besides the food for themselves, most farmers also grow some cash crops, products that can be sold for money such as cocoa, peanuts, rubber and palm oil. The cash crops often grow together with the food crops such as millet and maize. This method of growing different crops on the same field, called mixed cultivation, is better for the soil than when only one crop is grown, a method called monoculture. In mixed cultivation, all the plants are more resistant to pests and plant diseases and there is less risk of losing everything if there is a problem with one crop. On the other hand, the amounts produced per hectare are relatively low, especially since most Nigerian farmers only use very simple tools such as hoes and digging sticks.

Monoculture – a cocoa plantation

Preparing soil for mixed cultivation

Large plantations of 10,000 hectares of land or more account for about 10 per cent of Nigeria's agriculture. Products such as rice and wheat are grown on plantations for the home market. Other crops such as rubber and cocoa are only grown for export. Because a plantation has a lot of land and specializes in only one crop, much of the work can be done by machines and the plantation is run by a manager like a factory. This kind of farming needs a lot of capital to invest in technology as well as in fertilizer, pesticides and herbicides, which can also be dangerous to the environment. In this system of monoculture, it is possible to have high production, but on the other hand there are risks involved, especially when a crop is grown only for export and the farmers are completely dependent on the prices paid by the foreign market.

Nigeria has been exporting agricultural products, especially palm oil, to Europe for a very long time. In fact, in the 1800s the Niger River Delta was called the "Oil River" because so much palm oil came from there. The Europeans needed the palm oil to lubricate the machines of their new factories. Today palm oil, made by pressing the oil out of the fruit of the oil palm tree, is still exported from Nigeria. It is used to make many different products, including soap, cosmetics, margarine and medicine.

Another Kind of Oil Brings Changes

A key date in Nigerian history is 1956. That is when a different kind of oil was first found in Nigeria, not the kind from a plant, but from under the ground – petroleum. Suddenly it became possible for the country to earn much more money by exporting crude oil than by exporting agricultural products. The government began to invest a lot of money in oil production and new industries. The state gave very little support to the small farmers, but instead tried to increase the production of the large plantations. Without any state support and because they received only low prices for their products, the small farmers did not produce as much as they could have. At the same time, many young people went to the cities to look for jobs in the new factories, so there were fewer people to help on the small farms. Agricultural production went down at the same time that the population was growing rapidly. As a consequence, Nigeria had to import more and more food even though there was enough good farmland and a lot of unemployed people. In some years, Nigeria has had to spend up to 20 per cent of its oil income to import food. Nigeria used to export more agricultural products than it imported, but now it is just the opposite.

When the price of oil on the world market dropped in the 1980s, Nigeria's oil income was cut nearly in half. One way the government tried to deal with the big economic problems was to change its agricultural policy. They raised the guaranteed price paid to smaller farmers and also gave them loans, cheaper fertilizer, and agricultural advice. Because of these actions, agricultural production has risen again, but Nigeria still has to import large amounts of food in order to feed its people.

PRACTISE YOUR ENGLISH

These expressions can be used when explaining the import/export facts given in the graphs below:

Between 19... and 19... , food production / the amount of food produced per person decreased/increased rapidly/slowly.
The percentage of food that was imported increased / decreased / went from x% to y%.
In relation to the total amount of exports, agricultural products accounted for (much) more than / less than / about the same as crude oil in 19... .
In 19... agricultural products / crude oil accounted for all / nearly all / about half / about a fourth / about three-fourths / very little / none of Nigeria's total exports.

GEOGRAPHY TASKS

1 **Compare the main characteristics of a small traditional farm and a plantation. Organize your information in the form of a network.**

2 **In the last few decades the amount of monoculture in Nigeria has increased. Find and list three advantages and three disadvantages of monoculture. On balance, do you think monoculture is a good farming system? Explain.**

3 **Give short definitions of the terms *subsistence farming* and *cash crops*. Why are cash crops important to the individual farmer and to the country as a whole?**

4 **Using the information in Figures 2 and 3, describe and explain the development of Nigerian agriculture.**

5 **Why is it dangerous for Nigeria to rely too heavily on the export of a single product such as crude oil?**

Figure 2 **Food production and imports**

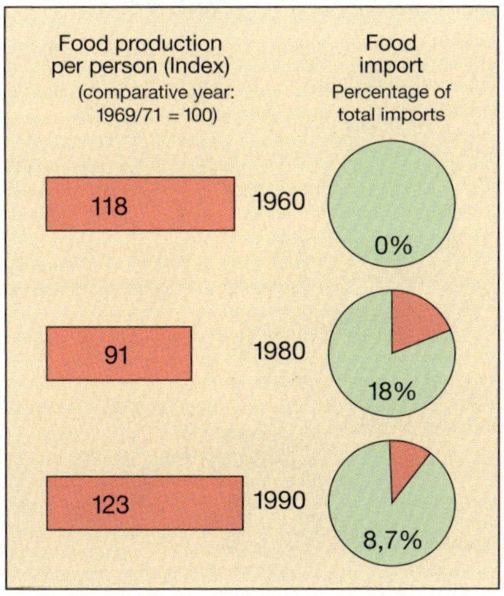

Figure 3 **Export of agricultural products and crude oil**

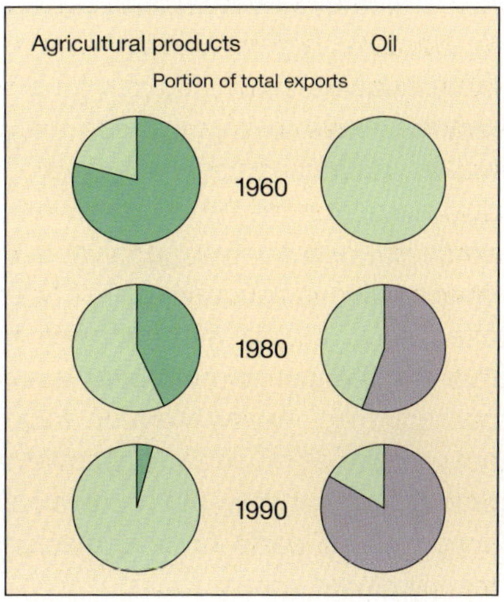

MODERN INDUSTRY: OPPORTUNITIES AND PROBLEMS

Because of its oil resources, Nigeria has a source of wealth that most developing countries can only dream of. Oil income could give a country a chance to build up modern industry, create jobs, and make it less economically dependent on other countries. But even though Nigeria was the world's fifth largest oil exporter in 1994, it was at the same time one of the 15 poorest countries in the world. Nigeria does have some modern industry, but in 1990 industry gave jobs to only 10 per cent of the population. What happened to the dream of transforming Nigeria into a wealthy country with a strong and growing industry?

Besides problems of corruption and mismanagement, a major structural problem was that the government first encouraged heavy industry such as iron and steel works as well as oil refineries and chemical plants. But because there were few factories to make things such as machines and consumer goods out of the iron and steel, much of it could not be used within the country and had to be exported in an unprocessed form. This is less profitable for a country than exporting manufactured goods. It also means that the country has to import many manufactured goods that it cannot make itself. There is also a lack of factories to process Nigeria's agricultural products such as palm oil, cocoa and rubber so they also must be exported as raw materials or in only a partly-processed form.

Because Nigeria does not manufacture large industrial machines and tools itself, its factories have to import these things. That makes production costs very high. There is also a lack of skilled workers, so the finished products are not only expensive, but often of poor quality. Products made in Nigeria can often not compete with imported products that are cheaper and of better quality. As a consequence, many Nigerian factories do not produce as much as they could. Another problem is that they often cannot afford to import spare parts for the machines and manufactured goods.

CREATING JOBS – AND UNEMPLOYMENT

Low industrial production is one reason why relatively few Nigerians have jobs in industry. Another reason is that when Nigerians and foreign investors did build factories, they used foreign technology to make them as modern as possible with robots and automation. That means that the new factories did not need so many workers. The new technology also meant that many people who had previously worked in small family businesses and traditional crafts lost their jobs because of competition with the factory-made products.

The government is trying to prevent unemployment problems by supporting small businesses that can offer jobs to many people. Brick-making, for example, gives jobs to many people when the work is done by hand. The relatively simple tools needed can also be made within the country and this in turn gives jobs to the tool-makers.

Through state employment programs, the government has already created over 200,000 jobs. Another goal is to distribute industry more evenly throughout the country and develop more industry in the central areas. In this way the coastal cities will not be so crowded. At present, more than half of all industrial jobs are in Lagos, a city with over 10 million people.

Baking Bread in Nigeria

Traditionally, Nigerian bread is baked in ovens made of rough stones and heated with charcoal. The loaves are kneaded by hand and put into the oven with a long wooden paddle. Five people working in a traditional bakery can make 400 loaves of bread in one day.

In 1984 a new bread factory was built in Kano by European specialists who had modern, high-tech machines flown in from Europe. This factory employed 60 people and could produce more than 50,000 loaves of bread a day at a price much lower than the traditional bakeries. Customers began buying the cheaper, factory-made bread. The small independent bakers tried to save money by dismissing some of their workers, but in the end, nearly 100 small bakeries in Kano had to close down because they could not compete with the modern factory.

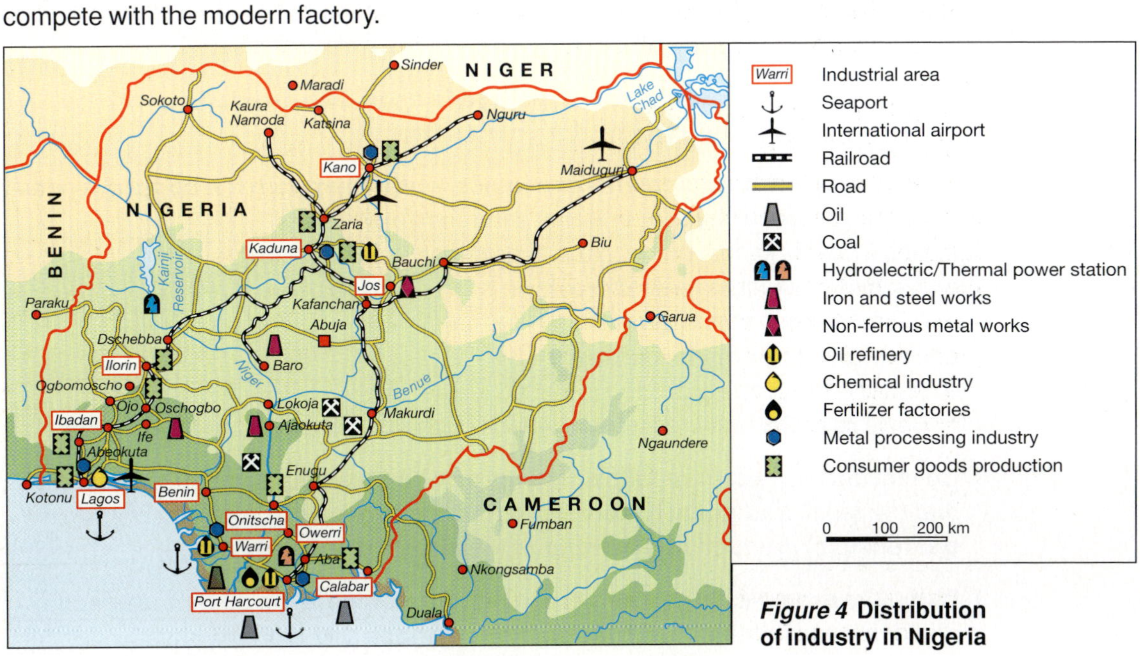

Figure 4 **Distribution of industry in Nigeria**

PRACTISE YOUR ENGLISH

The following expressions will help you when describing a map with the distribution of industries in a country:

This symbol stands for / represents ...
The oil refineries are (located) near/in/at ...
Chemical industries / fertilizer factories are concentrated in the ...
Coal / crude oil is found along the coast / near ... / in the central part ...

GEOGRAPHY TASKS

1 Match the heads and tails below to make complete sentences.

Heads	Tails
1 The government spent its income from oil on ...	a ... and use robots rather than employing people.
2 NIgeria's iron and steel has to be exported because ...	b ... so these products are exported in their unprocessed, and less valuable, state.
3 Exporting steel brings in less income for Nigeria than ...	c ... heavy industry such as iron and steel works, oil refineries and chemical plants.
4 The new heavy industries are highly automated ...	d ... it would get from exporting manufactured goods.
5 The government did not invest money in building factories to process palm oil, cocoa and rubber ...	e ... too few factories in the country are producing goods with it.

2 Describe the location of heavy manufacturing industries in Figure 4.

3 In Figure 4, what do you notice about the location of consumer goods factories compared with the location of heavy industry in Nigeria?

4 Explain why Nigeria is one of the world's poorest countries even though it has so much valuable crude oil.

5 What sort of industries do you think Nigeria should be developing with its income from oil? Give reasons for your answer.

6 Place the following statements in a logical order to explain the consequences of building a bread factory in Kano. Discuss your results.

a Small bakers cannot produce bread as cheaply as the factory
b Nearly 100 small bakeries close down
c Customers buy factory bread
d Price of factory bread is cheaper
e Bread factory is built; produces 50,000 loaves per day
f Small bakers dismiss workers to reduce costs and cut the price of their bread

8 Collect and compare the following information about the Kano bread factory and the small traditional bakers: (a) number of workers (b) number of loaves produced (c) productivity (number of loaves per person per day).

9 Using the example of the Kano bread factory, compare the advantages and disadvantages of traditional production methods and modern ones.

......The Restless Earth

THE STRUCTURE OF THE EARTH

The Earth can be compared to a giant peach. The "skin" is the relatively thin layer called the crust which completely covers the Earth. In some places under the ocean it is only about 5 km thick, under some mountains it can be over 100 km thick. The deeper parts of the crust have a temperature of up to 1200 °C. Although the world's deepest borehole only goes 12 km down into the Earth's crust, geologists believe that under the crust there is a layer called the mantle, which ranges in temperature from 1200–5000 °C. It is made of semi-molten rock that moves very slowly as the hotter rock rises and the cooler rock sinks. The centre of the Earth, called the core, is over 5000 °C, but is under so much pressure that it is a solid mass.

Figure 1 **The Earth's structure**

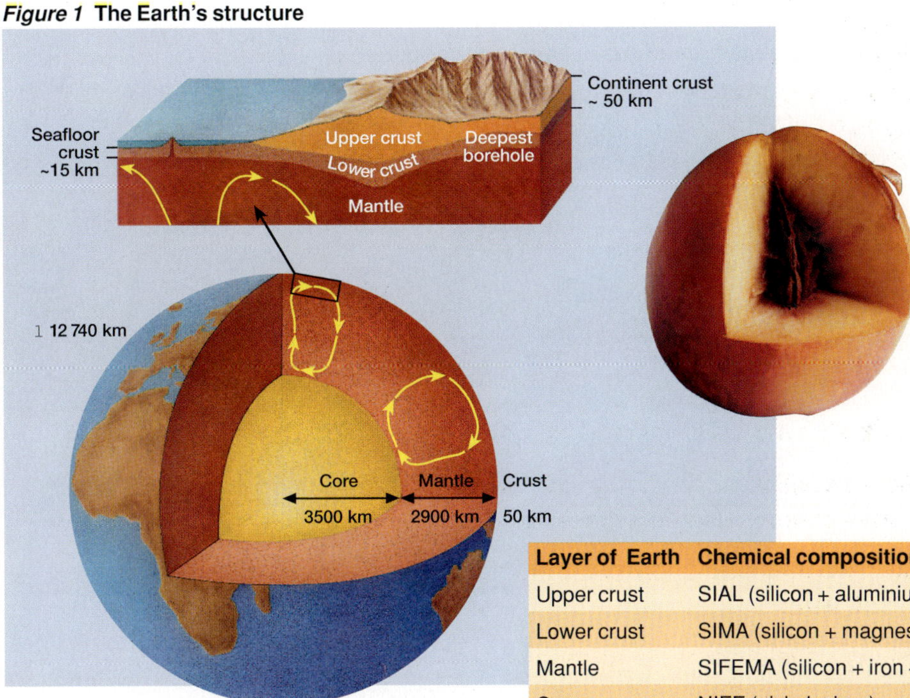

Layer of Earth	Chemical composition	Density (g/cm^3)
Upper crust	SIAL (silicon + aluminium)	2.7
Lower crust	SIMA (silicon + magnesium)	3.0
Mantle	SIFEMA (silicon + iron + magnesium)	5.0
Core	NIFE (nickel + iron)	10.0

GEOGRAPHY TASK ■■■

Copy the table shown below. Fill it in using the following words and expressions (some may be used more than once):

about 2900 km • solid • 1200°–5000 °C • semi-molten • 5–100 km
up to 1200 °C • about 3500 km • over 5000 °C

Layer	Thickness/depth	State of material	Temperature
Crust			
Mantle			
Core			

CONTINENTS ON THE MOVE

It is easy to notice that South America and Africa look like two giant puzzle pieces that have separated from each other. But continents are so big and solid, they cannot just move apart. Or can they? The German geographer Alfred Wegener (1880–1930) created a sensation in 1912 with his theory that all the continents were actually once a single land mass, which he named Pangaea or "All Land". Over a period of several million years, large pieces of Pangaea broke apart and moved around to form the continents we recognize today.

The theory that explains how continents can move is called plate tectonics. The crust of the Earth is cracked into about 15 big pieces called plates. Some of these plates carry continents or parts of continents, others carry parts of the seafloor. Because the inside of the Earth is so hot, convection currents cause molten magma in the Earth's mantle to move around. This movement puts pressure on the plates, which then move a few centimetres each year, about the same speed that fingernails grow. North America and Europe, for example, are still moving apart at the rate of about two centimetres a year. This process is happening so slowly that we only notice the movement when it causes an earthquake or the eruption of a volcano.

Figure 2 **Tectonic plates / earthquake and volcano zones**

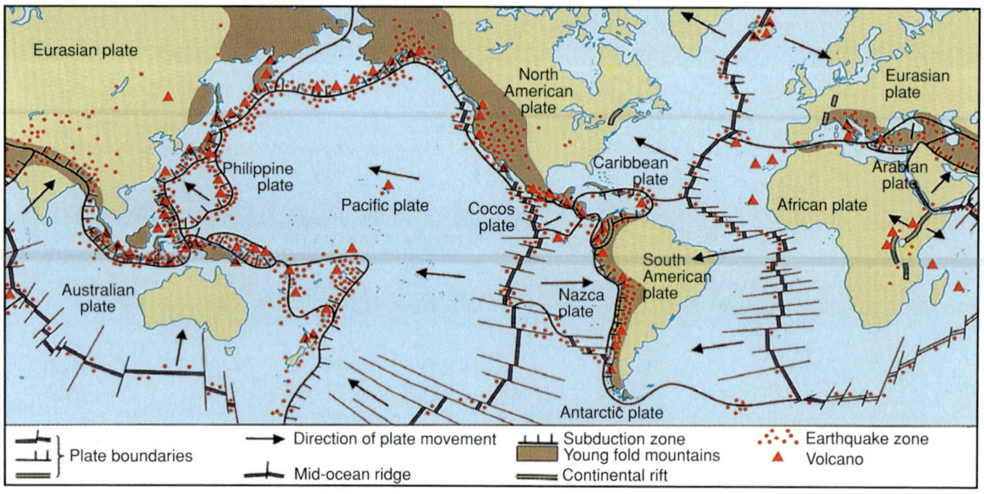

Three different things can happen along the cracks or plate boundaries. Sometimes two plates rub against each other as they move in opposite directions. A well-known example of this is the plate boundary that causes earthquakes in California. Sometimes plates pull apart from each other, usually under the ocean. This happens when magma from under the crust comes up and spreads like toothpaste to create a ridge of new crust under the ocean. This new crust causes the two plates to be pushed farther apart in opposite directions. A third type of activity happens when two plates bump into each other. Usually, one of the plates is pushed under the other one in a process called subduction, which means "to lead under". The material that is pushed under is melted by the heat deeper down in the Earth and becomes magma. The pressure created by subduction causes earthquakes and volcanoes to occur along these plate boundaries. When one of the plates involved in subduction carries land, this plate may be pushed up, forming fold mountains. The Andes in South America are one example of this process. A deep ocean trench is formed at the same time when the other plate is forced downwards and melted. When both plates carry land, the result can also be dramatic: the Ural Mountains between Russia and Europe as well as the Himalayas were created when two plates bumped into each other. The "crash", however, took millions of years.

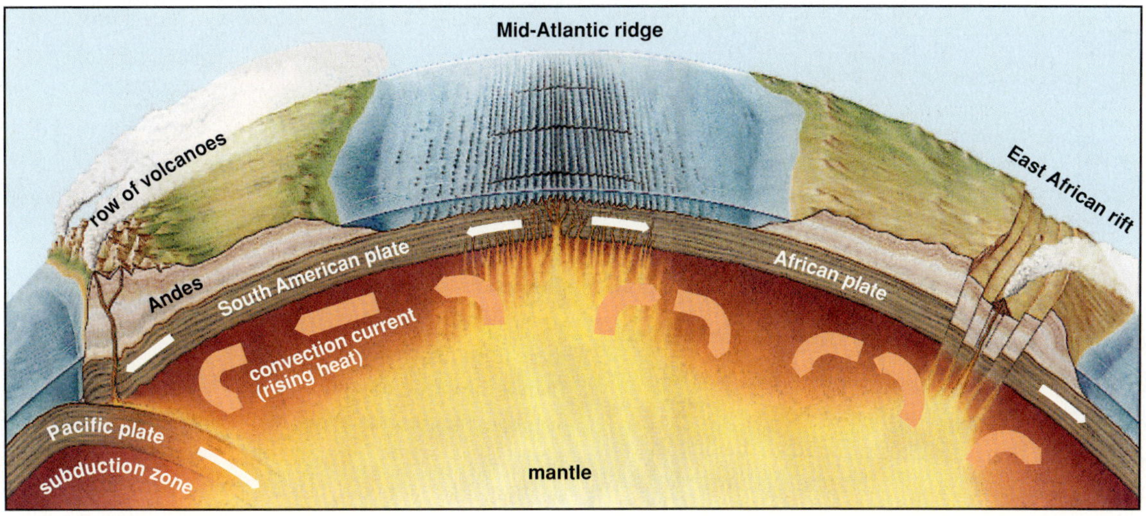

Figure 3 **Movement inside the Earth**

PRACTISE YOUR ENGLISH

You can use the following expressions when discussing plate tectonics:
The ... plate is moving toward the north/west/southwest.
The ... plate and the ... plate are drifting closer together / farther apart.

Subduction will occur when the ... plate collides with the ... plate.

The oceanic crust is sliding / dipping / being pushed under the continental crust.
Most earthquakes occur/happen along the plate boundaries.

GEOGRAPHY TASKS

1 Name and locate neighbouring pairs of plates on a world map. For each pair, find out in which direction each plate is moving and suggest the result of this movement.

2 Find the geographic terms to match the following descriptions:
a) At this place, new crust is created and the plates are constantly moving apart.
b) This is a region where the oceanic crust dips under the continental crust.
c) These two dramatic formations result from the upward and downward movement of plates.

3 Using a world map as a reference, draw the contours of the individual continents on a flat piece of styrofoam and cut them out using a knife or razor blade. First arrange the continents to form Pangaea, then rearrange them to look as they do today. Discuss the movements of the continents.

CALIFORNIA SUN *Magazine's special*

EARTHQUAKE AWARENESS

SUPPLEMENT

Don't wait till it's too late. Learn now what to do when an earthquake happens.

If indoors: Take cover under a heavy piece of furniture or against an inside wall and hold on. The most dangerous thing to do during the shaking of an earthquake is to try to leave the building. So stay inside!

What's Happening?

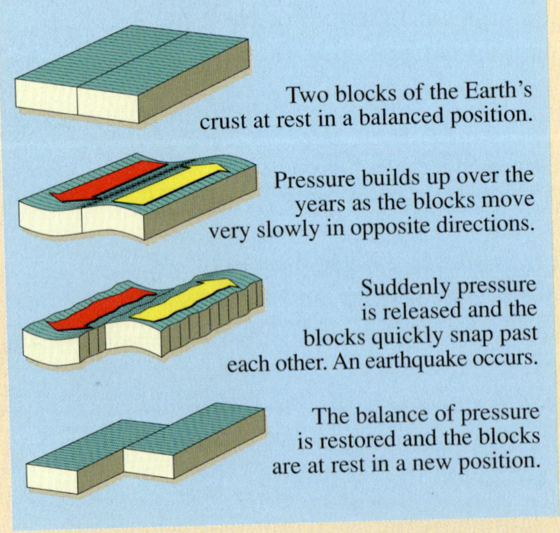

If outdoors: Move into the open, away from buildings, street lights and utility wires. Once in the open, stay there until the shaking completely stops.

If in a moving vehicle: Stop quickly and stay in the vehicle. Stay as far away as possible from bridges, overpasses, utility wires, trees or buildings. Once the shaking has stopped, proceed with caution. Avoid bridges or ramps.

Be prepared. Get your disaster supplies ready NOW. You should have:

a flashlight and extra batteries, a portable radio and extra batteries, a first aid kit and handbook, emergency food and water for at least 3 days, a nonelectric can-opener, supplies of medications you or your family are taking, cash and credit cards, and sturdy shoes.

Frequently Asked Questions

Is California going to fall into the Pacific Ocean?
No. In spite of all the jokes about people who move to Arizona to have a house at the beach when California falls into the ocean, it won't happen. On the other hand, Los Angeles is in fact moving north and in about 15 million years might be a suburb of San Francisco.

How do earthquakes happen?
The crust of the Earth under California is like a cracked egg with broken pieces of rock pushing against each other in different directions. Sometimes they move smoothly, sometimes they get stuck. When this happens, pressure builds up until the two pieces snap past each other quickly, like a stuck drawer that suddenly snaps open.

Can earthquakes be prevented?
No. Earthquakes are simply part of nature that happen frequently. In fact, it's earthquakes that have formed California's mountains and coastline. It is possible to minimize damage done by earthquakes by making buildings safer. After all, earthquakes don't kill people, falling buildings do.

The Earthquake Capital of California

"Be here when it happens" is the town's motto painted on the water tower. Parkfield, California (population 34) calls itself the earthquake capital of California. The tiny community in the hilly ranch country between Los Angeles and San Francisco is located right on the San Andreas fault and it experiences a major earthquake about every 23 years.

Margi Miller and her quake-warning system

It was the ringing of a Bavarian cowbell that woke Margi Miller, a 58-year-old rancher from Parkfield. She looked at her alarm clock: 4:31 am.

"The walls of my house were vibrating and the cups and plates were rattling in the kitchen," reported the blond widow. "After half a minute everything was over."

Margi jumped out of her bed, turned off the electricity and stayed under the door frame of her kitchen. If there isn't another trembling within 15 minutes, then this wasn't the "Big One", the big quake everyone in California is waiting for. At least that is what the people in Parkfield believe. And they should know: here in Parkfield the ground shakes nearly every day.

When Margi turned on the TV later she saw pictures of houses that were destroyed and burning and people who were in tears. An earthquake with a magnitude of 6.6 on the Richter scale had hit the city of Los Angeles. Close to tears herself, Margi looks at her private warning system, a souvenir from Oberammergau. "This bell from a Bavarian cow is always right."

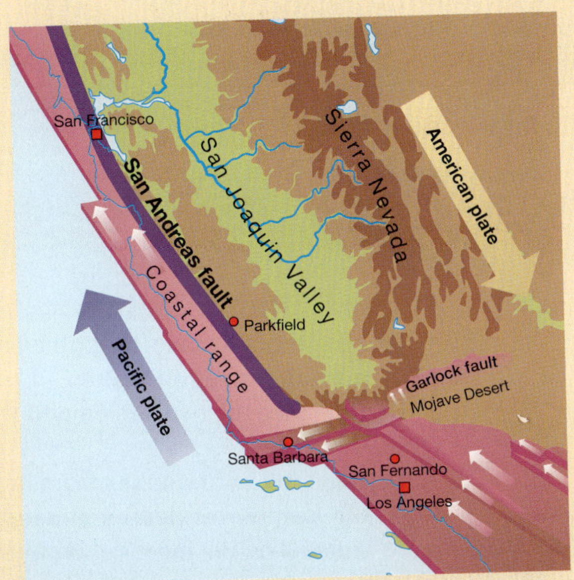

The Richter scale, created by American geologist Charles Richter in 1935, measures the size of the ground waves generated by an earthquake.

Richter scale:	**1–2.5**	Generally not felt, but recorded on seismometers; 900,000 per year
	3.5	Felt by people nearby; 50,000 per year
	4.5	Some damage may occur; 1000 per year
	6.0	Destructive earthquake; 120 per year
	7.0	Major earthquake; about 10 each year
	8.0	Great earthquake; once every 5–10 years
	9.0 or more	One or two per century

GEOGRAPHY TASKS ■■■

1 Give reasons for each of the rules about what to do during an earthquake. What else do you think it is important to know about what to do before or after an earthquake?

2 Pretend you are a television reporter in Parkfield, California. Interview Margi Miller about her experiences.

3 The following table gives information about some major earthquakes of the twentieth century. With the help of an atlas, plot the places on a map. In which three large regions have earthquakes taken place most frequently? Compare this information with the map showing the tectonic plates in Figure 2.

Figure 4 Details of selected major earthquakes

Year	Place	Magnitude	Number of deaths	Latitude	Longitude
1906	San Francisco, CA (USA)	8.2	700	38 °N	122 °W
1923	Tokyo (Japan)	8.3	143,000	36 °N	140 °W
1960	Agadir (Morocco)	5.9	12,000	30 °N	10 °W
1970	Chimbotc (Peru)	7.8	66,800	9 °S	78 °W
1976	Guatemala City	7.9	22,000	15 °N	90 °W
1976	Tangshan (China)	7.6	655,000	40 °N	118 °E
1976	Udine (Italy)	6.9	900	46 °N	13 °E
1977	Bucharest (Romania)	7.2	1200	44 °N	26 °E
1985	Mexico City	7.8	4000	20 °N	100 °W
1989	Spitak (USSR)	6.9	25,000	41 °N	44 °E
1989	Loma Prieta, CA (USA)	7.1	67	37 °N	121 °W
1990	Northern Iran	7.7	50,000	39 °N	48 °E

4 The Mercalli scale, though less popular than the Richter scale, is also used for measuring earthquakes. Collect information about the Mercalli scale. Compare the two scales and their advantages and disadvantages.

5 Rank the earthquakes from largest to smallest magnitude and add to your list the number of deaths caused by earthquakes. Do the most powerful earthquakes cause the most deaths? Suggest reasons for your answer.

6 Collect newspaper articles as well as information from the Internet about earthquakes and display them together with a world map. Mark the earthquake locations on the map and compare the locations with a plate tectonics map.

THE ROCK CYCLE

We think of rock as a symbol of strength and stability and use expressions like "as solid as a rock", but actually rocks are always changing. Most of these changes, however, take several million years.

The cycle of rocks begins about 60 km inside the Earth in the hot magma layer where molten rock moves around slowly like honey. When a volcano erupts, some of this hot molten rock breaks through the Earth's crust. Volcanic rock is formed when lava from a volcano cools into small crystals. This is called extrusive igneous rock. "Igneous" comes from the Latin word for "fire". Basalt and obsidian are two examples of volcanic rock.

Another kind of igneous rock is formed when the hot magma does not break through the Earth's crust, but cools and becomes solid under the surface. This is called intrusive rock. Because it cools very slowly, its minerals form large crystals. Granite is one kind of intrusive rock that is found under many mountains. In some places, we can see granite when the overlying surface wears away.

Rocks are changed by weathering and erosion. Wind, temperature changes, water and ice cause pieces of rock to break off into smaller and smaller pieces. Rivers carry the pieces into lakes and oceans. When this material finally stops moving, it "sits" or settles down and becomes what is called sediment. As more and more layers "sit" on top of each other, the weight of the new layers presses the material closer and closer together and a solid mass called sedimentary rock is formed. Layers of sand pressed together become sandstone, clay becomes shale, and limy mud becomes limestone. Dead plants and animals can also be pressed into the layers of rock and, with time, become fossils, which can sometimes be seen. In the Alps, people have found fossils of old sea creatures which lived a long time ago when the rock which formed the Alps was under a great ocean.

Metamorphic rock, a third type of rock, gets its name from the Greek word meaning "change of form". Both igneous and sedimentary rock can be changed into metamorphic rock through the action of heat, pressure and water on the arrangement of minerals. Gneiss (pronounced "nice"), for example, was once granite. Marble was once limestone. Movements of the Earth's crust can cause pieces of rock to fall into deeper, hotter areas and be changed there. These movements can also lead to rocks being pushed together, causing folds. The Alps were formed in just this way.

Figure 5 **The rock cycle**

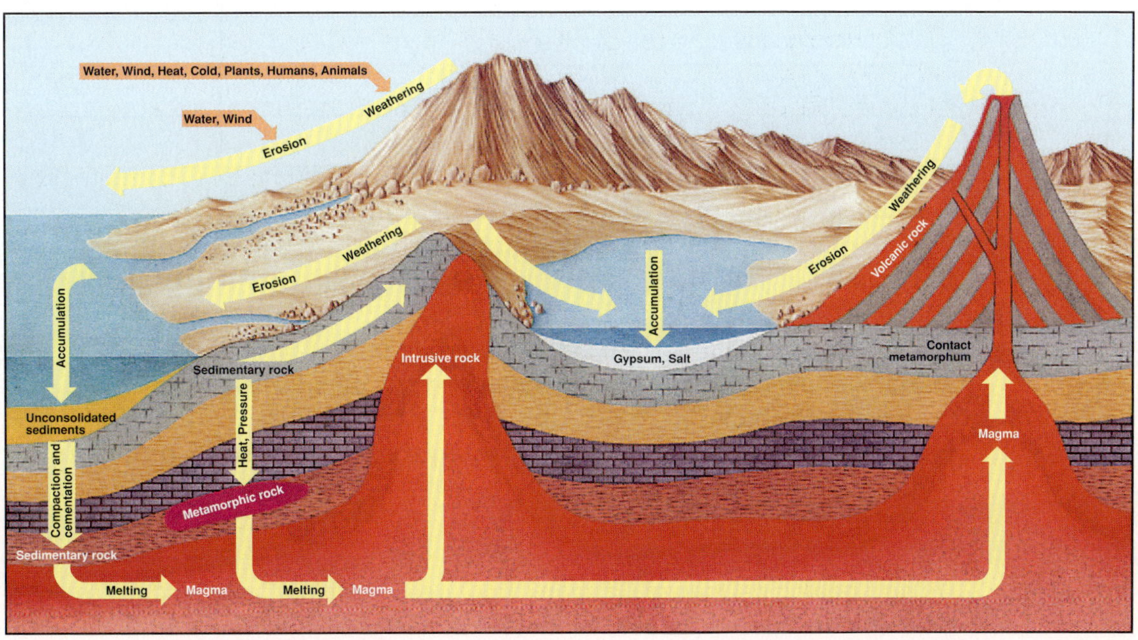

Fossilization

Sandstone formations

Volcanic rock

Granite

GEOGRAPHY TASKS ■■■

1 a) **Copy the table below and arrange the following words and phrases in it. Some boxes in the table will have more than one entry in them.**
gneiss • crystals rearranged • granite • sandstone • marble • may contain fossils • basalt • layered • limestone • large crystals • small crystals • magma cools slowly beneath the surface • great heat and pressure causes rocks to change • magma cools rapidly on the surface • particles of mud and sand are carried to the sea where they settle

Type of rock	Formation process	Rock structure	Named example
Volcanic igneous			
Intrusive igneous			
Sedimentary			
Metamorphic			

b) **Describe in your own words the formation process of each type of rock.**

2 **Using the information on the rock cycle in Figure 5**
a) **explain how a particle from an igneous rock could one day form part of a sedimentary rock.**
b) **explain how a particle from a sedimentary rock could one day form part of an extrusive igneous rock.**
c) **explain why we call this a rock cycle.**

3 **Using a geological map of Germany, find areas in which these kinds of rock predominate: granite, basalt, sandstone, limestone.**

Vocabulary / Geographical Dictionary (English – German)

Note: The relevant geographical terms which are found in this book are listed in **blue.**

A

abortion	Abtreibung
abroad	im Ausland
absenteeism	häufige Abwesenheit
"absentee" owner	ortfremder Eigentümer
abundance [ə'bʌndəns]	Reichtum, Fülle
acid rain	saurer Regen
(to) **adapt to sth.**	anpassen an etwas
agribusiness ['ægrəbɪznɪs]	industrielle Landwirtschaft
agriculture ['ægrəkʌltʃə]	Landwirtschaft, Ackerbau (und Viehzucht)
alfalfa	Luzerne, Alfalfa
almond ['ɑːmənd]	Mandel
altitude ['æltɪtjuːd]	Höhe
ancestor ['ænsestə]	Vorfahr, Ahne
approximate [ə'prɒksɪmət]	ungefähr
aqueduct ['ækwɪdʌkt]	Aquädukt
area ['eəriə]	Gebiet, Gegend
farming ~	Anbaugebiet
industrial ~	Industriegebiet
metropolitan ~ [ˌmetrə'pɒlɪtən]	Großstadtgebiet
recreation ~ [ˌrekri'eɪʃn]	Erholungsbebiet
residential ~	Wohngebiet
rural ~ ['rʊərəl]	ländliche Gegend
(to) **ascend** [ə'send]	aufsteigen
assembly line [ə'sembli]	Fließband
association	Vereinigung
(to) **attract sb./sth.**	jn./etwas anziehen

B

bargain ['bɑːgən]	günstiges Angebot
basalt	Basalt
basin ['beɪsn]	Becken
barrel ['bærəl]	Fass; Tonne; Barrel (= Maßeinheit für Erdöl, ca. 159 l)
belly-up	mit dem Bauch nach oben
belt	Gürtel
(to) **benefit from sth.**	von etw. profitieren
benefits	Vorteile, Begünstigungen
biotope ['baɪətəʊp]	Biotop
birch [bɜːtʃ]	Birke
(to) **blend in**	sich einfügen
boom	Boom, Aufschwung
brand	Marke
brick-making	das Herstellen von Backsteinen
broadleaf forest ['brɔːdliːf]	Laubwald
business vitality [vaɪ'tæləti]	lebendiges Geschäftsleben

C

canyon ['kænjən]	Schlucht
cattle	(Rind-)Vieh
central authority	zentrale Staatsgewalt
channel	Kanal
charcoal	Holzkohle
checkerboard	Schachbrett
chemical plant ['kemɪkəl]	Chemiefabrik
chlorofluorocarbon [ˌklɔːrə'flɔːrəkɑːbən]	Fluorchlorkohlenwasserstoff
clay	Tonerde, Lehm
climate ['klaɪmət]	Klima
~ graph	Klimadiagramm
moderate ~ ['mɑdərət]	gemäßigtes Klima
coal [kəʊl]	Kohle
coastal plain	Küstenebene
cold-storage depot	Kühlhaus
combine ['kɒmbaɪn]	Mähdrescher
commune ['kɒmjuːn]	(Volks)Kommune, Gemeinde
compensation [ˌkɒmpen'seɪʃn]	Entschädigung
competitor [kəm'petɪtə]	Konkurrent
concern [kən'sɜːn]	Besorgnis, Sorge
coniferous forest [ˌkəʊ'nɪfərəs]	Nadelwald
(to) **conquer sb./sth.** ['kɒŋkə]	jn./etwas erobern, bezwingen
consumer goods	Konsumgüter
consumption [kən'sʌmpʃn]	Konsum, Verbrauch
energy ~	Energieverbrauch
continental rift	kontinentaler Grabenbruch
(to) **contribute**	beitragen
contraceptive [kɒntrə'septiv]	Empfängnisverhütungsmittel
convection current	Konvektionsstrom
copper	Kupfer
core	Kern
corn	Mais
~-on-the-cob	Maiskolben
corporation	Handelsgesellschaft, Privatunternehmen, Gesellschaft mit beschränkter Haftung
cost	Kosten
cotton	Baumwolle
(to) **cough** [kɒf]	husten
(to) **cram**	pauken, büffeln

credit	Kredit
crop	Getreideart, (Feld-)Frucht
cash ~	kommerzieller Anbau
food ~	Anbau für die eigene Versorgung
(to) **create sth.**	etwas schaffen
cross-section map	Querschnittskarte
crude oil	Rohöl
crust	Kruste
cube [kju:b]	Würfel
custom harvesting	Ernte auf Bestellung

D

dacha	Datscha
dairy	Milchwirtschaft
decade ['dekeɪd]	Jahrzehnt
deep ocean trench	Tiefseegraben
defence industry	Verteidigungsindustrie
demand	Nachfrage *(Wirtschaft)*
demographic data	Bevölkerungsdaten
(to) **demonstrate sth.**	etwas zeigen, beweisen
density-shaded map	thematische Karte, die Bevölkerungsdichte durch unterschiedliche Schattierungen beschreibt
(to) **descend** [dɪ'send]	her-, hinuntergehen
destructive [dɪ'strʌktɪv]	zerstörerisch
(to) **dig**	graben
dike	Damm, Deich
(to) **dip under sth.**	unter etwas untertauchen
disaster	Katastrophe
(to) **dismiss sb.**	jn. entlassen
distribution [dɪstrɪ'bjuʃn]	Verbreitung, Verteilung
~ sector	Vertriebsbereich
diverse [daɪ'vɜ:s]	verschieden(artig), unterschiedlich
door frame	Türrahmen
drawer [drɔ:]	Schublade
(to) **dump sth.**	etwas auskippen, abladen
dust	Staub

E

eager	eifrig
earthquake	Erdbeben
economic	wirtschaftlich, Wirtschafts-
~ growth	Wirtschaftswachstum
economy	Wirtschaft
centrally planned ~	zentrale Planwirtschaft
free-market ~	freie Marktwirtschaft
efficient [ɪ'fɪʃnt]	tüchtig, leistungsfähig
effort ['efət]	Bemühung, Versuch
emergency [i'mɜ:dʒənsi]	Notfall
emphasis ['emfəsɪs]	Akzent, Betonung
employee	Arbeitnehmer(in), Angestellte(r)
employment	Anstellung, Arbeit, Beschäftigung
~ structure	Arbeitsstruktur
(to) **encourage sth.**	etwas fördern, unterstützen
(to) **enforce sth.**	etwas vollstrecken, geltend machen
entrepreneurial [ˌɒntrəprə'nɜ:riəl]	unternehmerisch
equipment	Ausrüstung
erosion [ɪ'rəʊʒn]	Erosion, Auswaschung
eruption [ɪ'rʌpʃn]	Ausbruch
(to) **estimate** ['estɪmeɪt]	schätzen
ethnic	ethnisch, Volks-
expense	(Un)Kosten
(to) **export sth.**	etwas exportieren, ausführen

F

fabric ['fæbrɪk]	Stoff
facade [fə'sɑ:d]	Fassade
(to) **face sb./sth.**	etwas angehen, jm. mutig entgegentreten
farm	Bauernhof, Gut(shof), Farm
collective ~	Kolchose
~ commodity [kə'mɒdəti]	landwirtschaftliches Erzeugnis
~ production input	Investition für die Landwirtschaft
state ~	Staatsfarm, Sowchose
fault line [fɔ:lt]	Verwerfungslinie
feature ['fi:tʃə]	Merkmal
federation	Zusammenschluss, Föderation
feedlot	(eingezäunter) Mastplatz für die intensive (Rind-)Viehhaltung, feedlot
fenced	ein-, umgezäunt
fertile ['fɜ:taɪl]	fruchtbar, ertragreich
fertility	Fruchtbarkeit, Ergiebigkeit
fertilizer ['fɜ:tɪlaɪzə]	Dünger
financial backing [faɪ'nænʃl]	finanzielle Unterstützung
fish-breeding	Fischzucht
(to) **float** [fləʊt]	treiben
(to) **flood** [flʌd]	überschwemmen, -fluten
flow chart [fləʊ]	Ablaufdiagramm
focus	Schwerpunkt, Brennpunkt
(to) **focus**	sich konzentrieren auf
fold	Falte, Falz
~ mountains	Faltengebirge
food-processing	Nahrungsmittelkonservierung, -verarbeitung
foreign debt ['fɒrən det]	Auslandsschulden
forestry	Forstwirtschaft
fossil ['fɒsl]	Fossil
~ fuel ['fju:əl]	fossiler Brennstoff
freeway	Autobahn *(USA)*

G

gap	Kluft, Lücke
garden plot	Beet im Garten
(to) **generate sth.**	etwas erzeugen
genetic technology [dʒəˈnetɪk]	Gentechnologie
global market	Weltmarkt
global warming	Erwärmung der Erdatmosphäre
GNP = gross national product	*(siehe unten)*
goal	Ziel
goat	Ziege
grain	Getreide
~ **farm**	Getreidefarm
feed ~	Futter
gravel [ˈgrævl]	Kies
greasy [ˈgri:si]	schmierig, ölbeschmiert
greenhouse operation	Betreiben eines Gewächshauses
gross domestic product [grəʊs]	Bruttoinlandsprodukt
gross national product	Bruttosozialprodukt
ground	Boden
growing period	Vegetationszeit

H

handicapped	behindert
haze [heɪz]	Dunstschleier, Nebel
herbicide [hɜ:bɪsaɪd]	Unkrautvernichtungsmittel
high-tech [ˌhaɪ ˈtek]	Spitzentechnologisch-, Hitech-
(to) **hike**	wandern
(to) **hire sb.**	jn. anstellen
hoe	Hacke
hog [hɒg]	(Mast)schwein
holding company	Dach-, Holdinggesellschaft
hot spot	attraktive Industriestandort
humid [ˈhju:mɪd]	feucht
hut [hʌt]	Hütte

I

(to) **import sth.**	etwas einführen, importieren
import tariff [ˈtærɪf]	Importzoll
impressive	beeindruckend
improvement	Verbesserung, Besserung, Erweiterung
industrial region [ˈri:dʒən]	industrialisierte Gegend
industrialized country	Industrieland
industry	Industrie
heavy ~	Schwerindustrie
light ~	Leichtindustrie
incentive	Anreiz
income	Einkommen
inefficient [ɪnɪˈfɪʃnt]	unwirtschaftlich
inhabitant	Bewohner, Einwohner
(to) **invest in sth.**	in etwas investieren, Kapital in Sachgütern anlegen
investment	Investition
investor	Investor
iron ore [ˈaɪən ɔ:]	Eisenerz
(to) **irrigate**	bewässern
irrigation [ɪrɪˈgeɪʃn]	Bewässerung

J

joint venture [ˈventʃə]	Gemeinschaftsunternehmen
joint-stock company	Aktiengesellschaft

K

key	Schlüssel
(to) **knead sth.** [ni:d]	etwas kneten
knowledge-intensive	wissensintensiv

L

label [ˈleɪbl]	Etikett
labour [ˈleɪbə]	Arbeit
~**-intensive**	arbeitsintensiv
land	Land, Boden
~**mark**	Wahrzeichen
~ **ownership**	Grundbesitz(erschaft), Grundeigentum(srecht)
~ **reclamation** [ˌrekləˈmeɪʃn]	Landgewinnung
~**slide**	Erdrutsch
latitude [ˈlætɪtju:d]	Breitengrad
line of ~	Breitengradslinie
lava	Lava
layer	Schicht, Lage
leadership	Führung
(to) **lease sth.** [li:s]	etwas pachten
less developed country	Entwicklungsland
(to) **level sth.**	etwas einebnen, planieren
liberalisation	Liberalisierung
lichen [ˈlaɪkɪn]	Flechte
life expectancy [ɪkˈspektənsi]	Lebenserwartung
lime [laɪm]	Linde
limestone	Kalkstein
limy mud [ˈlaɪmi]	kalkhaltiger Schlamm
line graph	Kurvendiagramm
loan	Darlehen
loaf (pl. **loaves**)	Laib
(to) **locate sb./sth.**	jn./etwas ausfindig machen
loess [ˈləʊes]	Löss
location [ləʊˈkeɪʃn]	Lage
(to) **long for sth.**	sich sehnen nach etwas
lowlands	Tief-, Flachland
loyalty	Loyalität
(to) **lubricate sth.**	etwas gleitfähig machen, schmieren, ölen

M

magma	Magma
magnitude	Größe, Größenordnung
(to) **maintain sth.**	etwas aufrecht erhalten, beibehalten
(to) **make sth. worthwhile**	etwas lohnend machen
maize [meɪz]	Mais
mantle	(Erd-)Mantel
manufacturing plant	Fabrik
marble	Marmor
market	Markt
domestic/home ~	Binnenmarkt
marketing company	Marketinggesellschaft
meadow ['medəʊ]	Wiese
medieval [ˌmedi'i:vl]	mittelalterlich
megalopolis [ˌmegə'lɒpələs]	Megalopolis, (aus mehreren Städten bestehende) Riesenstadt
memory circuit board ['meməri 'sɜ:kɪt bɔ:d]	Speicherplatine
mid-ocean ridge [rɪdʒ]	mittelozeanischer Rücken
millet	Hirse
migrant worker ['maɪgrənt]	Wanderarbeiter
mine	Bergwerk, Grube, Zeche
mineral oil	Mineralöl
mismanagement	Misswirtschaft
mixed cultivation	Mischkultur
mixed forest	Mischwald
(to) **monitor sth.** ['mɒnɪtə]	etwas überwachen, kontrollieren
monoculture ['mɒnəʊkʌltʃə]	Monokultur
monopoly [mə'nɒpəli]	Monopol
moss [mɒs]	Moos

N

natural gas	Erdgas
nomad ['nəʊmæd]	Nomade, Nomadin
non-ferrous metal [ˌnɒn'ferəs 'metəl]	nicht eisenhaltiges Metal
nursery product ['nɜ:səri]	Gärtnerei-, Baumschulenprodukt

O

oak [əʊk]	Eiche
obsidian [ɒb'sɪdiən]	kieselsäurereiches, glasiges Gestein, Obsidian
occasional [ə'keɪʒənl]	gelegentlich
(to) **occur** [ə'kɜ:]	sich ereignen, vorkommen, geschehen
origin ['ɒrɪdʒɪn]	Ursprung, Herkunft
ostrich ['ɒstrɪtʃ]	(Vogel)Strauß
outline sketch	Umrißzeichnung, -skizze
ozone layer ['əʊzəʊn]	Ozonschicht

P

packing plant	Schlachthof
paddle	*(hier)* Brotschieber
Pangaea [pænd'ʒi:ə]	Pangäa, Urkontinent
pattern	Muster, Schema
peach	Pfirsich
peasant ['peznt]	Kleinbauer
per capita [ˌpə 'kæpɪtə]	pro-Kopf-
permafrost	Dauerfrost
permanent	ständig
pest	Schädling
pesticide ['pestɪsaɪd]	Schädlingsbekämpfungsmittel, Pestizid
petrochemical	petrochemisch
~ product	petrochemisches Erzeugnis
phlegm [flem]	Schleim
pie graph	Tortendiagramm
plain	Ebene
plantation	Plantage, Pflanzung
plaque [plɑ:k]	Schild, Tafel
plate	Platte
~ boundary ['baʊndri]	Plattengrenze
~ tectonics	Plattentektonik
Antarctic ~	antarktische Platte
plateau [plæ'təʊ]	Plateau, Hochebene
pole	Pfeiler
pollution [pə'lu:ʃn]	(Umwelt)Verschmutzung
air ~	Luftverschmutzung
water ~	Wasserverschmutzung
population [ˌpɒpju'leɪʃn]	Bevölkerung
~ density	Bevölkerungsdichte
~ growth	Bevölkerungswachstum
~ movement	Bevölkerungswanderung
~ pyramid ['pɪrəmɪd]	Bevölkerungspyramid
populous ['pɒpjələs]	dichtbesiedelt, -bevölkert
porcelain ['pɔ:səlɪn]	Porzellan
(to) **pour sth.** [pɔ:]	etwas schütten, gießen
poultry ['pəʊltri]	Geflügel
power station	Kraftwerk
hydroelectric ~ ['haɪdrəʊɪ'lektrɪk]	Wasserkraftwerk
nuclear ~	Atomkraftwerk
precision optics [prɪ'sɪʃn]	feinoptische Geräte
(to) **predict sth.**	etwas vorhersagen
prediction	Vorhersage
(to) **predominate** [prɪ'dɒmɪneɪt]	vorherrschen
prefabricated house [pri:'fæbrɪkeɪtɪd]	Fertighaus
(to) **pretend** [pri:'tend]	vorgeben, so tun als ob
previous ['pri:viəs]	vorherig, vorhergehend, vorangegangen
privilege ['prɪvəlɪdʒ]	Privileg
(to) **process sth.**	*(technisch)* etwas verarbeiten

production	Produktion, Herstellung, Erzeugung
~ **costs**	Produktions-, Herstellungskosten
~ **level**	Produktionsleistung
~ **line**	Fließband, Fertigungsstraße
productive	produktiv
productivity	Produktivität
public transport system	öffentlicher Personennahverkehr (ÖPNV)

R

(to) **raise sth.**	etwas ziehen, anbauen (*Pflanzen)*
range [reɪndʒ]	Reichweite
(to) **rank**	einstufen, einordnen
raw material [ˌrɔː məˈtɪəriəl]	Rohstoff
(to) **reclaim land**	Land gewinnen
(to) **recognize sb./sth.**	jn./etwas erkennen, anerkennen
recreational [ˌrekriˈeɪʃənl]	Freizeit-
recycling	Recycling, Wiederverwertung
(to) **reduce sth.**	etwas reduzieren, verkürzen, herabsetzen, einschränken
(to) **refer to sth.** [rɪˈfɜː]	nachschauen in etwas, etwas konsultieren
refinery [rɪˈfaɪnəri]	Raffinerie
~ **product**	Raffinerieprodukt
oil ~	Ölraffinerie
reindeer [ˈreɪndɪə]	Rentier
reliable [rɪˈlaiəbl]	verlässlich
relief [rɪˈliːf]	Oberflächenform der Erde, Relief
~ **payment**	Hilfeleistung, Unterstützungszahlung
repetitive [rəˈpetɪtɪv]	sich dauernd wiederholend
(to) **require sth.**	etwas brauchen, benötigen
research [rɪˈsɜːtʃ, ˈriːsɜːtʃ]	Forschung
reservoir [ˈrezəvwɑː]	Stausee, Auffangbehälter
resources [ˈrɪsɔːsɪz]	Bodenschätze
natural ~	Bodenschätze
mineral ~	mineralische Bodenschätze
responsibility	Verantwortung
~ **system**	System, das Bauern ermöglicht, über ihre Produktion selbst zu entscheiden *(China)*
(to) **restore sth.**	etwas wiederherstellen
(to) **retire**	aufhören zu arbeiten, in den Ruhestand treten
retirement community	Gemeinschaft von Leuten im Rentenalter
ridge [rɪdʒ]	Rücken, Kamm
ripe	reif, zeitig
river oasis (pl. **oases**) [əʊˈeɪsɪs, -siːz]	Flussoase
rock	Fels, Gestein
~ **cycle** [ˈsaɪkl]	Kreislauf der Gesteine
extrusive igneous ~ [ɪkˈstruːsɪv ˈɪgniːəs]	an der Erdoberfläche erstarrtes Eruptivgestein
igneous ~	Eruptivgestein
intrusive ~	Intrusiv-, Tiefengestein
molten ~	geschmolzenes Gestein
metamorphic ~	metamorphes Gestein
sedimentary ~	Sedimentgestein
semi-molten ~	halb geschmolzenes Gestein
volcanic ~ [vɒlˈkænɪk]	Vulkangestein
rooster	Hahn
(to) **rule**	beherrschen, herrschen über
rubble	Schutt, Schotter
rusty	verrostet

S

sake [ˈsɑːkeɪ]	Reiswein
sandstone	Sandstein
sawmill	Sägewerk
scale	Skala
Mercalli ~	Mercalli Skala
Richter ~ [ˈrɪktə]	Richter Skala
(to) **scrap sth.**	etwas verschrotten, ausrangieren
sector	Sektor
primary ~ [ˈpraɪməri]	primärer Sektor
secondary ~	sekundärer Sektor
tertiary ~ [ˈtɜːʃəri]	tertiärer Sektor
secure [sɪˈkjʊə]	sicher, gesichert
sediment	Bodenablagerung
seed	Saatgut, Samen
~ **farm**	Saatgutfarm
seismometer [ˌsaɪzˈmɒmɪtə]	Seismometer, Erdbebenmesser
(to) **service sth.**	etwas warten
(to) **settle**	sich niederlassen
settlement	Ansiedlung, (Wohn-)Siedlung
shale	Schiefer
(to) **shame sb./oneself**	jn. beschämen, sich schämen
shareholder	Aktionärin, Aktionär
shift	Schicht
(to) **shift**	(sich) verlagern, verschieben
shipping lane	Schiffsweg
silk	Seide
~ **cloth** [klɒθ]	Stoff aus Seide
silt	Schlamm, Schlick
site	Standort, Stelle, Platz
sketch	Skizze
skilled worker	Facharbeiter
(to) **slaughter sth.** [ˈslɔːtə]	etwas schlachten
sledding	Schlitten fahren
(to) **slide**	sich schieben, sich gleiten

	lassen
slope	Hang
slum	Elendsviertel
smokestack	Schornstein
smooth [smu:ð]	glatt, ruhig
(to) **snap open**	aufschnappen
society [ˌsəˈsaɪəti]	die Gesellschaft
soil [sɔɪl]	Erde
soup kitchen	Armenküche
source [sɔ:s]	Quelle
soybean	Sojabohne
spare part	Ersatzteil
sparse	spärlich, dünn(gesät)
Special Economic Zone	Wirtschaftssonderzone *(China)*
(to) **specialize in sth.**	sich auf etwas spezialisieren
spice	Gewürz
spruce [spru:s]	Fichte
steelworks	Stahlwerk
stick	Stock
(to) **stick up**	hochstehen, herausschauen
stock	Wertpapiere, Aktien
stuck	stecken geblieben
sturdy	kräftig, robust
subduction [sʌbˈdʌkʃn]	Abtauchen *(einer Erdplatte)*
~ **zone**	Abtauchzone
subsidy [ˈsʌbzɪdi]	Subvention, Zuschuss
state ~	staatliche Subvention, Zuschuss
subsistence farmer [sʌbˈsɪstəns]	Bauer, der ausschließlich für den Eigenbedarf produziert
suburb [ˈsʌbɜ:b]	Vorort
(to) **be suitable for sth.** [ˈsu:təbl]	für etwas geeignet, passend sein
sulphur dioxide [ˈsʌlfə daɪˈɒksaɪd]	Schwefeldioxid
(to) **suffer from sth.**	unter etwas leiden
supply	Angebot *(Wirtschaft)*
surface [ˈsɜ:fɪs]	Oberfläche
swamp [swɒmp]	Sumpf, Morast
synthetic fibre [ˌsɪnˈθetɪk ˈfaɪbə]	Kunstfaser

T

tax	Steuer
~ **incentive**	Steueranreiz
~ **privilege** [ˈprɪvəlɪdʒ]	Steuervorteil
terrace [ˈterəs]	Terrasse, Geländestufe
(to) **thaw** [θɔ:]	tauen
thermal energy [ˈθɜ:ml]	Wärmekraft
~ **plant**	Wärmekraftwerk
(to) **threaten sb./sth.** [ˈθretn]	jn./etwas bedrohen
ties [taɪz]	Verbindungen
tool-maker	Werkzeughersteller
trade	Handel
~ **balance** [ˈbæləns]	Handelsbilanz
~ **deficit** [ˈdefɪsɪt]	Handelsdefizit
~ **surplus**	Handelsüberschuss
(to) **transact sth.**	etwas abwickeln, durchführen
transportation network	öffentliches Verkehrssystem
(to) **trap sth.**	etwas auffangen
treasure [ˈtreʒə]	Schatz, Reichtum
triangle [ˈtraɪeŋgl]	Dreieck
tribe	Stamm
tsunami	Tsunami, Flutwelle
typhoon	Taifun

U

unemployment	Arbeitslosigkeit
unfavourable	ungünstig
union [ˈju:niən]	Gewerkschaft
unit	Einheit
urban [ˈɜ:bən]	städtisch, Stadt-
~**isation** [ˌɜ:bənaɪˈzeɪʃn]	Verstädterung
utility wire [ju:ˈtɪləti ˈwaɪə]	Versorgungsleitung

V

vacation [veɪˈkeɪʃn]	Urlaub
value [ˈvælju:]	Wert
valuable [ˈvæljuəbl]	wertvoll
value-added product	technologisch hochwertiges Produkt
valueless [ˈvælju:les]	wertlos
victim	Opfer
virtual	virtuell
volcano [vɒlˈkeɪnəʊ]	Vulkan

W

wage	Lohn
(to) **waste sth.**	etwas verschwenden, vergeuden
watershed [ˈwɔ:təʃed]	Wasserscheide
wave	Welle
wealth [welθ]	Reichtum, Besitz, Vermögen
wealthy	reich, wohlhabend
(to) **wear sth. away**	etwas auswaschen, aushöhlen
weathering	Witterung
wheat [wi:t]	Weizen
wooden raft	hölzernes Floß
work force	Arbeiterschaft, Arbeitskräfte

Y

yak	Grunzochse, Yak

Geographical Dictionary (German–English)

A

Ablaufdiagramm flow chart
Abtauchen subduction *(tectonic plates)*
Ackerbau (und Viehzucht) agriculture, farming
Ackerland arable land
Aktiengesellschaft joint-stock company
Aktionärin, Aktionär shareholder
Anbau für die eigene Versorgung food crop
Angebot supply *(commerce)*
Angestellte, Angestellter employee, worker
Ansiedlung settlement
Anstellung employment
Aquädukt aqueduct
Arbeit labour, work, employment
  ~erschaft, ~skräfte work force
  ~nehmerin, ~nehmer employee, worker
  ~sintensiv labour-intensive
  ~slosigkeit unemployment
Atomkraftwerk nuclear power station
Auffangbehälter reservoir
Ausbruch eruption
ausführen (to) export
Auslandsschulden foreign debt
Auswaschung erosion
Autobahn freeway *(USA)*, motorway *(UK)*

B

Basalt basalt
Bauer farmer
  ~, der nur für den Eigenbedarf produziert subsistence farmer
Bauernhof farm
Baumwolle cotton
Becken basin
Bergwerk mine
Beschäftigung work, job, employment
Betreiben eines Gewächshauses greenhouse operation
Bevölkerung population
  ~sdaten demographic data
  ~sdichte population density
  ~spyramide population pyramid
  ~swachstum population growth
  ~swanderung population movement
bewässern (to) irrigate
Bewässerung irrigation
Bewohnerin, Bewohner inhabitant
Binnenmarkt domestic market, home market
Biotop biotope
Boden ground
  ~ablagerung sediment
Bodenschätze resources
  natürliche ~ natural resources
  mineralische ~ mineral resources
Breitengrad latitude
  ~slinie line of latitude
Bruttoinlandsprodukt gross domestic product
Bruttosozialprodukt gross national product (GNP)
Chemiefabrik chemical plant

D

Damm dike
Darlehen loan
Dauerfrost permafrost
dichtbesiedelt, -bevölkert populous
Dünger fertilizer

E

Ebene plain
ortfremder Eigentümer "absentee" owner
einführen (to) import
Einwohnerin, Einwohner inhabitant
Eisenerz iron ore
Elendsviertel slum
Energieverbrauch energy consumption
Entwicklungsland less developed country
Erdbeben earthquake
  ~messer seismometer
Erde soil, earth
Erdgas natural gas
(Erd-)Mantel mantle
Erdrutsch landslide
Erholungsgebiet recreation area
Erosion erosion
ertragreich *(Landwirtschaft)* fertile
Erwärmung der Erdatmosphäre global warming
Erzeugung production, manufacture
exportieren (to) export

F

Fabrik	manufacturing plant
Facharbeiter	skilled worker
Falte, Falz	fold
Faltengebirge	fold mountains
feedlot	feedlot
Fels	rock
Fichte	spruce
Flachland	lowlands
Flechte	lichen
Fließband	assembly line, production line
Fluorchlorkohlenwasserstoff	chlorofluorocarbonate
Flussoase	river oasis (pl. oases)
Flutwelle	tsunami, tidal wave
Föderation	federation
Forschung	research
Forstwirtschaft	forestry
Fossil	fossil
fruchtbar	fertile
Fruchtbarkeit	fertility
(eingezäunter) Futterplatz für die intensive (Rind-) Viehhaltung	feedlot

G

Gebiet	area, region, district, zone
Großstadt~	metropolitan area
Wohn~	residential area
Gegend	area, region, district, part
ländliche ~	rural area
Geländestufe	terrace
Gemeinde	commune, community
Gemeinschaft von Leuten im Rentenalter	retirement community
Gemeinschaftsunternehmen	joint venture
Gentechnologie	genetic technology
Gestein	rock
an der Erdoberfläche erstarrtes ~	extrusive igneous rock
Eruptiv~	igneous rock
geschmolzenes ~	molten rock
halb geschmolzenes ~	semi-molten rock
Intrusiv~	intrusive rock
kieselsäurereiches, glasiges ~	obsidian
Kreislauf der ~e	rock cycle
metamorphes ~	metamorphic rock
Sediment~	sedimentary rock
Getreide	grain
~farm	grain farm
Granit	granite
Größe, Größenordnung	magnitude *(earthquake)*
Grundbesitz(erschaft), Grundeigentum(srecht)	land ownership
Gut(shof)	(large) farm

H

Handel	trade
~sbilanz	trade balance
~sdefizit	trade deficit
~sgesellschaft	corporation, commercial company
~süberschuss	trade surplus
Hang	slope
Herstellung	production, manufacture
~skosten	production costs
Hirse	millet
Hochebene	plateau
Holzkohle	charcoal
Hütte	hut

I

importieren	(to) import
Importzoll	import tariff
industrialisierte Gegend	industrial region
Industrie	industry
~land	industrialized country
Leicht~	light industry
Schwer~	heavy industry
investieren	(to) invest
Investition	investment
~ für die Landwirtschaft	farm production input
Investor	investor

K

Kalkstein	limestone
kalkhaltiger Schlamm	limy mud
Kamm	ridge
Kanal	channel
Kern	core
Kies	gravel
Kleinbauer	peasant
Klima	climate
~diagramm	climate graph
gemäßigtes ~	moderate climate
Kohle	coal
Kolchose	collective farm
kommerzieller Anbau	cash crop
Kommune	commune
Konkurrent	competitor
Konsumgüter	consumer goods
kontinentaler Grabenbruch	continental rift
Konvektionsstrom	convection current
Kosten	cost
Kraftwerk	power station/plant
Wärme~	thermal energy plant
Wasser~	hydroelectric power station/plant
Atom~	nuclear power station/plant
Kredit	credit

Kruste	crust
Kühlhaus	cold-storage depot
Kupfer	copper
Kurvendiagramm	line graph
Küstenebene	coastal plain

L

Lage	layer *(geology)*
Land gewinnen	(to) reclaim land
Landgewinnung	land reclamation
Landwirtschaft	agriculture, farming
industrielle ~	agribusiness
landwirtschaftliches Erzeugnis	farm commodity
Laubwald	broadleaf forest
Lava	lava
Lebenserwartung	life expectancy
Lehm	clay
Löss	loess
Luftverschmutzung	air pollution

M

Mais	corn, maize
Magma	magma
Marketinggesellschaft	marketing company
Megalopolis	megalopolis
Merkmal	feature
Mineralöl	mineral oil
Mischkultur	mixed cultivation
Mischwald	mixed forest
Misswirtschaft	mismanagement
mittelozeanischer Rücken	mid-ocean ridge
Monokultur	monoculture
Monopol	monopoly
Moos	moss

N

Nachfrage	demand *(commerce)*
Nadelwald	coniferous forest
Nahrungsmittelkonservierung, -verarbeitung	food-processing
nicht eisenhaltiges Metal	non-ferrous metal
Nomade, Nomadin	nomad

O

Oberfläche	surface
Obsidian	obsidian
öffentlicher Personennahverkehr (ÖPNV)	public transport system, transportation network
Ozonschicht	ozone layer

P

Pangäa	Pangaea
Pestizid	pesticide
petrochemisch	petrochemical
~es Erzeugnis	petrochemical (product)
Pflanzung, Plantage	plantation
Plateau	plateau
Platte	plate
~ngrenze	plate boundary
~ntektonik	plate tectonics
antarktische ~	Antarctic plate
Produktion	production
~skosten	production costs
~sleistung	production level
produktiv	productive
~ität	productivity
pro-Kopf-	per capita

Q

Querschnittskarte	cross-section map

R

Raffinerie	refinery
~produkt	refinery product
Öl~	oil refinery
Recycling	recycling
(aus mehreren Städten bestehende) Riesenstadt	megalopolis
saurer Regen	acid rain
Relief	relief
Rentier	reindeer
Rohöl	crude oil
Rohstoff	raw material
Rücken	ridge *(mountains)*

S

Saatgut	seed
~farm	seed farm
Sägewerk	sawmill
Sandstein	sandstone
Schädling	pest
Schädlingsbekämpfungsmittel	pesticide
Schicht	layer *(geology)*
Schiefer	shale
Schiffsweg	shipping lane
Schlachthof	slaughterhouse, packing plant
Schlucht	canyon
Schotter, Schutt	rubble
Schwefeldioxid	sulfur dioxide
Seismometer	seismometer
Sektor	sector
primärer ~	primary sector
sekundärer ~	secondary sector
tertiärer ~	tertiary sector
Siedlung	settlement

Skala	scale
Mercalli ~	Mercalli scale
Richter ~	Richter scale
Sowchose, Staatsfarm	state farm
Stadt-, städtisch	urban
Stahlwerk	steelworks
Standort	site
Stausee	reservoir
Steuer	tax
~anreiz	tax incentive
~vorteil	tax privilege
Stoff	fabric
Subvention	subsidy
staatliche ~	state subsidy
System, das Bauern ermöglicht, über ihre Produktion selbst zu entscheiden	responsibility system *(China)*

T

tauen	(to) thaw
Taifun	typhoon
technologisch hochwertiges Produkt	value-added product
Terrasse	terrace
Tiefengestein	intrusive rock
Tiefland	lowlands
Tiefseegraben	deep ocean trench
Tonerde	clay
Tsunami	tsunami, tidal wave

U

überfluten, überschwemmen	(to) flood
Umweltverschmutzung	pollution
Unkrautvernichtungsmittel	herbicide

V

Vegetationszeit	growing period
verarbeiten	(to) process
Verstädterung	urbanisation
Vertriebsbereich	distribution sector
Verwerfungslinie	fault line
Vorort	suburb
Vulkan	volcano
~gestein	volcanic rock

W

Wahrzeichen	landmark
Wärmekraft	thermal energy
Wasserscheide	watershed
Wasserverschmutzung	water pollution
Weizen	wheat
Weltmarkt	global market
Werkzeughersteller	tool-maker
Wiederverwertung	recycling
Wirtschaft	economy
~ssonderzone	Special Economic Zone *(China)*
~swachstum	economic growth
freie Markt~	free-market economy
zentrale Plan~	centrally planned economy
wirtschaftlich, Wirtschafts-	economic
wissensintensiv	knowledge-intensive
Witterung	weathering
Wohnsiedlung	settlement

Z

Zeche	mine
Zusammenschluss	federation
Zuschuss	subsidy

Bildquellen

Action Press, Hamburg: S. 33 (oben), 46 (unten links), 50 (unten); Bavaria, Gauting: S. 10 (Stock Imagery), 11 (Stock Imagery); Bilderberg, Hamburg: S. 29 (© Burkard); Corbis, London: S. 16 (unten); DESWOS, Köln: S. 57 (rechts); Focus, Hamburg: S. 38 (unten), 46 (mitte rechts, © Urs Klyver; unten rechts, © Michael Nichols), 50 (oben, © Lindsay Hebberd), 52 (oben, © Greg Girard), 53 (oben links, © Catherine Henriette; oben mitte, © Michael Yamashita; oben rechts, © Martin Parr; unten rechts, © Greg Girard); Das Fotoarchiv, Essen: S. 32 (links, © Jochen Tack), 33 (unten, © Christopher Morris); Friedhoff, Dortmund: S. 47 (unten); Globus, Hamburg: S. 39; Hugo, Oberhausen: S. 32 (rechts), 47 (oben), 51 (oben); The Hutchison Library, London: S. 55, 58 (oben und unten, © Anna Tully), 60 (oben); Janicke, München: S. 47 (zweites von unten); Jürgens, Köln: S. 24 (zweites von oben, zweites von unten); Jürgens Ost- und Europa-Photo, Berlin: S. 31 (oben links); Klohn, Vechta: S. 7 (oben links, unten links); Mauritius, Mittenwald: S. 57 (links); Mayer, Hattingen: S. 43; Rachel Kibler-Melby, Berlin: S. 16 (oben); Oxford University Press, Oxford: S. 45 (oben und unten; aus M. Day/R. Grenyer/S. Chapman, *Oxford Geography Programme*, Vol. 2, OUP, 1997. By permission of Oxford University Press); Steenmans, Korschenbroich: S. 7 (oben rechts), 9, 18 (unten); Taubert, Springer: S. 60 (unten); TCIDA, Tuscaloosa: S. 17; Tony Stone, Hamburg: S. 2 (mitte rechts), 3 (oben links), 18 (oben links); Toyota Deutschland: S. 41; Werner + Pfleiderer, Stuttgart: S. 61 (oben); Wölfel, Tröstau: S. 70 (0ben links); ZEFA, Düsseldorf: S. 38 (oben, © Koch).